THE WORLD ACCORDING TO QUANTUM MECHANICS

Why the Laws of Physics Make
Perfect Sense After All

THE WORLD ACCORDING TO QUANTUM MECHANICS

Why the Laws of Physics Make Perfect Sense After All

Ulrich Mohrhoff

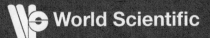

World Scientific

NEW JERSEY · LONDON · SINGAPORE · BEIJING · SHANGHAI · HONG KONG · TAIPEI · CHENNAI

Published by

World Scientific Publishing Co. Pte. Ltd.

5 Toh Tuck Link, Singapore 596224

USA office: 27 Warren Street, Suite 401-402, Hackensack, NJ 07601

UK office: 57 Shelton Street, Covent Garden, London WC2H 9HE

British Library Cataloguing-in-Publication Data
A catalogue record for this book is available from the British Library.

ISBN-13 978-981-4293-37-2
ISBN-10 981-4293-37-7

Printed in Singapore.

Preface

While still in high school, I learned that the tides act as a brake on the Earth's rotation, gradually slowing it down, and that the angular momentum lost by the rotating Earth is transferred to the Moon, causing it to slowly spiral outwards, away from Earth. I still vividly remember my puzzlement. How, by what mechanism or process, did angular momentum get transferred from Earth to the Moon? Just so Newton's contemporaries must have wondered at his theory of gravity. Newton's response is well known:

> I have not been able to discover the cause of those properties of gravity from phænomena, and I frame no hypotheses. ... to us it is enough, that gravity does really exist, and act according to the laws which we have explained, and abundantly serves to account for all the motions of the celestial bodies, and of our sea. [Newton (1729)]

In Newton's theory, gravitational effects were simultaneous with their causes. The time-delay between causes and effects in classical electrodynamics and in Einstein's theory of gravity made it seem possible for a while to explain "how Nature does it." One only had to transmogrify the algorithms that served to calculate the effects of given causes into physical processes by which causes produce their effects. This is how the electromagnetic field—a calculational tool—came to be thought of as a physical entity in its own right, which is locally acted upon by charges, which locally acts on charges, and which mediates the action of charges on charges by locally acting on itself.

Today this sleight of hand no longer works. While classical states are algorithms that assign trivial probabilities—either 0 or 1—to measurement outcomes (which is why they can be re-interpreted as collections of possessed properties and described without reference to "measurement"), quantum states are algorithms that assign probabilities *between* 0 and 1 (which is why they cannot be so described). And while the classical laws correlate measurement outcomes *deterministically* (which is why they can be interpreted in causal terms and thus as descriptive of physical processes), the quantum-mechanical laws correlate measurement outcomes *probabilistically* (which is why they cannot be so interpreted). In at least one respect, therefore, physics is back to where it was in Newton's time—and this with a vengeance. According to Dennis Dieks, Professor of the Foundations and Philosophy of the Natural Sciences at Utrecht University and Editor of *Studies in History and Philosophy of Modern Physics*,

> the outcome of foundational work in the last couple of decades has been that interpretations which try to accommodate classical intuitions are impossible, on the grounds that theories that incorporate such intuitions necessarily lead to empirical predictions which are at variance with the quantum mechanical predictions. [Dieks (1996)]

But, seriously, how could anyone have hoped to get away for good with passing off computational tools—mathematical symbols or equations—as physical entities or processes? Was it the hubristic desire to feel "potentially omniscient"—capable in principle of knowing the furniture of the universe and the laws by which this is governed?

If quantum mechanics is the fundamental theoretical framework of physics—and while there are a few doubters [e.g., Penrose (2005)], nobody has the slightest idea what an alternative framework consistent with the empirical data might look like—then the quantum formalism not only defies reification but also cannot be explained in terms of a "more fundamental" framework. We sometimes speak loosely of a theory as being more fundamental than another but, strictly speaking, "fundamental" has no comparative. This ·is another reason why we cannot hope to explain "*how* Nature does it." What remains possible is to explain "*why* Nature does it." When efficient causation fails, teleological explanation remains viable.

The question that will be centrally pursued in this book is: what does it take to have stable objects that "occupy space" while being composed of objects that do not "occupy space"?[1] And part of the answer at which we shall arrive is: quantum mechanics.

---

[1]The existence of such objects is a well-established fact. According to the well-tested theories of particle physics, which are collectively known as the Standard Model, the objects that do not "occupy space" are the quarks and the leptons.

As said, quantum states are algorithms that assign probabilities between 0 and 1. Think of them as computing machines: you enter (i) the actual outcome(s) and time(s) of one or several measurements, as well as (ii) the possible outcomes and the time of a subsequent measurement—and out pop the probabilities of these outcomes. Even though the time dependence of a quantum state is thus clearly a dependence on the times of measurements, it is generally interpreted—even in textbooks that strive to remain metaphysically uncommitted—as a dependence on "time itself," and thus as the time dependence of something that exists at every moment of time and evolves from earlier to later times. Hence the mother of all quantum-theoretical pseudo-questions: why does a quantum state have (or appear to have) two modes of evolution—continuous and predictable between measurements, discontinuous and unpredictable whenever a measurement is made?

The problem posed by the central role played by measurements in standard axiomatizations of quantum mechanics is known as the "measurement problem." Although the actual number of a quantum state's modes of evolution is zero, most attempts to solve the measurement problem aim at reducing the number of modes from two to one. As an anonymous referee once put it to me, "to solve this problem means to design an interpretation in which measurement processes are not different in principle from ordinary physical interactions." The way I see it, to solve the measurement problem means, on the contrary, to design an interpretation in which the central role played by measurements is *understood*, rather than swept under the rug.

An approach that rejects the very notion of quantum state evolution runs the risk of being dismissed as an ontologically sterile instrumentalism. Yet it is this notion, more than any other, that blocks our view of the ontological implications of quantum mechanics. One of these implications is that the spatiotemporal differentiation of the physical world is incomplete; it does not "go all the way down." The notion that quantum states evolve, on the other hand, implies that it does "go all the way down." This is not simply a case of one word against another, for the incomplete spatiotemporal differentiation of the physical world follows from the manner in which quantum mechanics assigns probabilities, which is *testable*, whereas the complete spatiotemporal differentiation of the physical world follows from an assumption about what is the case *between measurements*, and such an assumption is "not even wrong" in Wolfgang Pauli's famous phrase, inasmuch as it is neither verifiable nor falsifiable.

Understanding the central role played by measurements calls for a clear distinction between what measures and what is measured, and this in turn calls for a precise definition of the frequently misused and much maligned word "macroscopic." Since it is the incomplete differentiation of the physical world that makes such a definition possible, the central role played by measurements cannot be understood without dispelling the notion that quantum states evolve.

For at least twenty-five centuries, theorists—from metaphysicians to natural philosophers to physicists and philosophers of science—have tried to model reality from the bottom up, starting with an ultimate multiplicity and using concepts of composition and interaction as their basic explanatory tools. If the spatiotemporal differentiation of the physical world is incomplete, then the attempt to understand the world from the bottom up—whether on the basis of an intrinsically and completely differentiated space or spacetime, out of locally instantiated physical properties, or by aggregation, out of a multitude of individual substances—is doomed to failure. What quantum mechanics is trying to tell us is that reality is structured from the top down.

Having explained why interpretations that try to accommodate classical intuitions are impossible, Dieks goes on to say:

> However, this is a negative result that only provides us with a starting-point for what really has to be done: something conceptually new has to be found, different from what we are familiar with. It is clear that this constructive task is a particularly difficult one, in which huge barriers (partly of a psychological nature) have to be overcome. [Dieks (1996)]

Something conceptually new has been found, and is presented in this book. To make the presentation reasonably self-contained, and to make those already familiar with the subject aware of metaphysical prejudices they may have acquired in the process of studying it, the format is that of a textbook. To make the presentation accessible to a wider audience—not only students of physics and their teachers—the mathematical tools used are introduced along the way, to the point that the theoretical concepts used can be adequately grasped. In doing so, I tried to adhere to a principle that has been dubbed "Einstein's razor": everything should be made as simple as possible, but no simpler.

This textbook is based on a philosophically oriented course of contemporary physics I have been teaching for the last ten years at the Sri Aurobindo International Centre of Education (SAICE) in Puducherry (formerly Pondicherry), India. This non-compulsory course is open to higher secondary (standards 10–12) and undergraduate students, including students with negligible prior exposure to classical physics.[2]

The text is divided into three parts. After a short introduction to probability, Part 1 ("Overview") follows two routes that lead to the Schrödinger equation—the historical route and Feynman's path-integral approach. On the first route we stop once to gather the needed mathematical tools, and on the second route we stop once for an introduction to the special theory of relativity.

The first chapter of Part 2 ("A Closer Look") derives the mathematical formalism of quantum mechanics from the existence of "ordinary" objects—stable objects that "occupy space" while being composed of objects that do not "occupy space." The next two chapters are concerned with what happens if the objective fuzziness that "fluffs out" matter is ignored. (What happens is that the quantum-mechanical correlation laws degenerate into the dynamical laws of classical physics.) The remainder of Part 2 covers a number of conceptually challenging experiments and theoretical results, along with more conventional topics.

Part 3 ("Making Sense") deals with the ontological implications of the formalism of quantum mechanics. The penultimate chapter argues that quantum mechanics—whose validity is required for the existence of "ordinary" objects—in turn requires for its consistency the validity of both the Standard Model and the general theory of relativity, at least as effective theories. The final chapter hazards an answer to the question of *why* stable objects that "occupy space" are composed of objects that do not "occupy space." It is followed by an appendix containing solutions or hints for some of the problems provided in the text.

---

[2]I consider this a plus. In the first section of his brilliant Caltech lectures [Feynman *et al.* (1963)], Richard Feynman raised a question of concern to every physics teacher: "Should we teach the *correct* but unfamiliar law with its strange and difficult conceptual ideas ...? Or should we first teach the simple ... law, which is only approximate, but does not involve such difficult ideas? The first is more exciting, more wonderful, and more fun, but the second is easier to get at first, and is a first step to a real understanding of the second idea." With all due respect to one of the greatest physicists of the 20th Century, I cannot bring myself to agree. How can the second approach be a step to a real understanding of the correct law if *"philosophically we are completely wrong* with the approximate law," as Feynman himself emphasized in the immediately preceding paragraph? To first teach laws that are completely wrong philosophically cannot but impart a conceptual framework that eventually stands in the way of understanding the correct laws. The damage done by imparting philosophically wrong ideas to young students is not easily repaired.

I wish to thank the SAICE for the opportunity to teach this experimental course in "quantum philosophy" and my students—the "guinea pigs"—for their valuable feedback.

Ulrich Mohrhoff
August 15, 2010

Contents

PART 1
Overview

Chapter 1

Probability:
Basic concepts and theorems

The mathematical formalism of quantum mechanics is a probability calculus. The probability algorithms it places at our disposal—state vectors, wave functions, density matrices, statistical operators—all serve the same purpose, which is to calculate the probabilities of measurement outcomes. That's reason enough to begin by putting together what we already know and what we need to know about probabilities.

1.1 The principle of indifference

Probability is a measure of likelihood ranging from 0 to 1. If an event has a probability equal to 1, it is certain that it will happen; if it has a probability equal to 0, it is certain that it will not happen; and if it has a probability equal to 1/2, then it is as likely as not that it will happen.

Tossing a fair coin yields heads with probability 1/2. Casting a fair die yields any given natural number between 1 and 6 with probability 1/6. These are just two examples of the *principle of indifference*, which states:

If there are n mutually exclusive and jointly exhaustive possibilities (or possible events), and if we have no reason to consider any one of them more likely than any other, then each possibility should be assigned a probability equal to 1/n.

Saying that events are *mutually exclusive* is the same as saying that at most one of them happens. Saying that events are *jointly exhaustive* is the same as saying that at least one of them happens.

3

1.2   Subjective probabilities versus objective probabilities

There are two kinds of situations in which we may have no reason to consider one possibility more likely than another. In situations of the first kind, there are objective matters of fact that would make it certain, *if we knew them*, that a particular event will happen, but we don't know any of the relevant matters of fact. The probabilities we assign in this case, or whenever we know some but not all relevant facts, are in an obvious sense *subjective*. They are *ignorance* probabilities. They have everything to do with our (lack of) *knowledge* of relevant facts, but nothing with the *existence* of relevant facts. Therefore they are also known as *epistemic* probabilities.

In situations of the second kind, there are no objective matters of fact that would make it certain that a particular event will happen. There may not even be objective matters of fact that would make it more likely that one event will occur rather than another. There isn't any relevant fact that we are ignorant of. The probabilities we assign in this case are neither subjective nor epistemic. They deserve to be considered *objective*. Quantum-mechanical probabilities are essentially of this kind.

Until the advent of quantum mechanics, all probabilities were thought to be subjective. This had two unfortunate consequences. The first is that probabilities came to be thought of as something *intrinsically* subjective. The second is that something that was not a probability at all—namely, a *relative frequency*—came to be called an "objective probability."

1.3   Relative frequencies

Relative frequencies are useful in that they allow us to measure the likelihood of possible events, at least approximately, provided that trials can be repeated under conditions that are identical in all relevant respects. We obviously cannot measure the likelihood of heads by tossing a single coin. But since we can toss a coin any number of times, we can count the number N_H of heads and the number N_T of tails obtained in N tosses and calculate the fraction $f_N^H = N_H/N$ of heads and the fraction $f_N^T = N_T/N$ of tails. And we can expect the difference $|N_H - N_T|$ to increase significantly slower than the sum $N = N_H + N_T$, so that

$$\lim_{N \to \infty} \frac{|N_H - N_T|}{N_H + N_T} = \lim_{N \to \infty} |f_N^H - f_N^T| = 0\,. \qquad (1.1)$$

In other words, we can expect the relative frequencies f_N^H and f_N^T to tend to the probabilities p_H of heads and p_T of tails, respectively:

$$p_H = \lim_{N \to \infty} \frac{N_H}{N}, \qquad p_T = \lim_{N \to \infty} \frac{N_T}{N}. \tag{1.2}$$

1.4 Adding and multiplying probabilities

Suppose you roll a (six-sided) die. And suppose you win if you throw either a 1 or a 6 (no matter which). Since there are six equiprobable outcomes, two of which cause you to win, your chances of winning are $2/6$. In this example it is appropriate to *add* probabilities:

$$p(1 \vee 6) = p(1) + p(6) = \frac{1}{6} + \frac{1}{6} = \frac{1}{3}. \tag{1.3}$$

The symbol $\vee$ means "or." The general rule is this:

Sum rule. Let $\mathcal{W}$ be a set of w mutually exclusive and jointly exhaustive events (for instance, the possible outcomes of a measurement), and let $\mathcal{U}$ be a subset of $\mathcal{W}$ containing a smaller number u of events: $\mathcal{U} \subset \mathcal{W}$, $u < w$. The probability $p(\mathcal{U})$ that one of the events $e_1, \ldots, e_u$ in $\mathcal{U}$ takes place (no matter which) is the sum $p_1 + \cdots + p_u$ of the respective probabilities of these events.

One nice thing about relative frequencies is that they make a rule such as this virtually self-evident. To demonstrate this, let N be the total number of trials—think coin tosses or measurements. Let N_k be the total number of trials with outcome e_k, and let $N(\mathcal{U})$ be the total number of trials with an outcome in $\mathcal{U}$. As N tends to infinity, N_k/N tends to p_k and $N(\mathcal{U})/N$ tends to $p(\mathcal{U})$. But

$$\frac{N(\mathcal{U})}{N} = \frac{N_1 + \cdots + N_u}{N} = \frac{N_1}{N} + \cdots + \frac{N_u}{N}, \tag{1.4}$$

and in the limit $N \to \infty$ this becomes

$$p(\mathcal{U}) = p_1 + \cdots + p_u. \tag{1.5}$$

Suppose now that you roll two dice. And suppose that you win if your total equals 12. Since there are now 6×6 equiprobable outcomes, only one of which causes you to win, your chances of winning are $1/(6 \times 6)$. In this example it is appropriate to *multiply* probabilities:

$$p(6 \wedge 6) = p(6) \times p(6) = \frac{1}{6} \times \frac{1}{6} = \frac{1}{36}. \tag{1.6}$$

The symbol $\wedge$ means "and." Here is the general rule:

Product rule. The *joint probability* $p(e_1 \wedge \cdots \wedge e_v)$ of v *independent* events $e_1, \ldots, e_v$ (that is, the probability with which *all* of them happen) is the product of the probabilities $p(e_1), \ldots, p(e_v)$ of the individual events.

It must be stressed that the product rule only applies to independent events. Saying that two events a, b are *independent* is the same as saying that the probability of a is independent of whether or not b happens, and *vice versa*.

As an illustration of the product rule for two independent events, let $a_1, \ldots, a_J$ be mutually exclusive and jointly exhaustive events (think of the possible outcomes of a measurement of a variable A), and let $p_1^a, \ldots, p_J^a$ be the corresponding probabilities. Let $b_1, \ldots, b_K$ be a second such set of events with corresponding probabilities $p_1^b, \ldots, p_K^b$. Now draw a 1×1 square with coordinates x, y ranging from 0 to 1. Partition it horizontally into J strips of respective width p_j^a. Partition it vertically into K strips of respective width p_k^b. You now have a square partitioned into $J \times K$ rectangles with respective areas $p_j^a \times p_k^b$. Since a joint measurement of A and B is equivalent to throwing a dart in such a way that it hits a random position (x, y) within the square, the joint probability $p(a_j \wedge b_k)$ equals the corresponding area.

Problem 1.1. *We have seen that the probability of obtaining a total of 12 when rolling a pair of dice is 1/36. What is the probability of obtaining a total of (a) 11, (b) 10, (c) 9?*

Problem 1.2. $(*)^1$ *In 1999, Sally Clark was convicted of murdering her first two babies, which died in their sleep of sudden infant death syndrome. She was sent to prison to serve two life sentences for murder, essentially on the testimony of an "expert" who told the jury it was too improbable that two children in one family would die of this rare syndrome, which has a probability of 1/8,500. After over three years in prison, and five years of fighting in the legal system, Sally was cleared by a Court of Appeal, and another two and a half years later, the "expert" pediatrician Sir Roy Meadow was found guilty of serious professional misconduct. Amazingly, during the trial nobody raise the objection that an expert pediatrician was not likely to be an expert statistician. Meadow had argued that the probability of two sudden infant deaths in the same family was $(1/8, 500) \times (1/8, 500) = 1/72, 250, 000$. Explain why he was so terribly wrong.*

---

[1] A star indicates that a solution or a hint is provided in Appendix A.

1.5   Conditional probabilities and correlations

If the events a_j and b_k are *not* independent, we must distinguish between *marginal probabilities*, which are assigned to the possible outcomes of either measurement without taking account of the outcome of the other measurement, and *conditional probabilities*, which are assigned to the possible outcomes of either measurement depending on the outcome of the other measurement. If a_j and b_k are not independent, their joint probability is

$$p(a_j \wedge b_k) = p(b_k|a_j)\,p(a_j) = p(a_j|b_k)\,p(b_k)\,, \tag{1.7}$$

where $p(a_j)$ and $p(b_k)$ are marginal probabilities, while $p(b_k|a_j)$ is the probability of b_k conditional on the outcome a_j and $p(a_j|b_k)$ is the probability of a_j conditional on the outcome b_k. This gives us the useful relation

$$p(b|a) = \frac{p(a \wedge b)}{p(a)}\,. \tag{1.8}$$

Another useful rule is

$$p(a) = p(a|b)\,p(b) + p(a|\bar{b})\,p(\bar{b})\,, \tag{1.9}$$

where b and $\bar{b}$ are two mutually exclusive and jointly exhaustive events. (To obtain $\bar{b}$ is to obtain any outcome other than b.) The validity of this rule is again readily established with the help of relative frequencies. We obviously have that

$$\frac{N(a)}{N} = \frac{N(a \wedge b)}{N} + \frac{N(a \wedge \bar{b})}{N} = \frac{N(a \wedge b)}{N(b)}\frac{N(b)}{N} + \frac{N(a \wedge \bar{b})}{N(\bar{b})}\frac{N(\bar{b})}{N}\,, \tag{1.10}$$

where N is the number of joint measurements of two variables, one with the possible outcome a and one with the possible outcome b. In the limit $N \to \infty$, $N(a)/N$ (the left-hand side of Eq. 1.10) tends to the marginal probability $p(a)$, while the right-hand side of this equation tends to the right-hand side of Eq. (1.9), as will be obvious from a glance at Eq. (1.8).

An important concept is that of (probabilistic) correlation. Two events a, b are *correlated* just in case that $p(a|b) \neq p(a|\bar{b})$. Specifically, a and b are *positively* correlated if $p(a|b) > p(a|\bar{b})$, and they are *negatively* correlated if $p(a|b) < p(a|\bar{b})$. Saying that a and b are *independent* is thus the same as saying that they are *uncorrelated*, in which case $p(a|b) = p(a|\bar{b}) = p(a)$.

Problem 1.3. (∗) *Let's Make a Deal was a famous game show hosted by Monty Hall. In it a player was to open one of three doors. Behind one door there was the Grand Prize (for example, a car). Behind the other doors* *there were booby prizes (say, goats). After the player had chosen a door, the host opened a different door, revealing a goat, and offered the player the opportunity of choosing the other closed door. Should the player accept the offer or should he stick with his first choice? Does it make a difference?*

Problem 1.4. (∗) *Which of the following statements do you think is true? (i) Event A happens more frequently because it is more likely. (ii) Event A is more likely because it happens more frequently.*

Problem 1.5. (∗) *Suppose we have a 99% accurate test for a certain disease. And suppose that a person picked at random from the population tests postive. What is the probability that this person actually has the disease?*

1.6 Expectation value and standard deviation

Another two important concepts associated with a probability distribution are the *expected/expectation value* (or *mean*) and the *standard deviation* (or *root mean square deviation from the mean*).

The expected value associated with the measurement of an observable with K possible outcomes v_k and corresponding probabilities $p(v_k)$ is

$$\langle v \rangle \overset{\text{Def}}{=} \sum_{k=1}^{K} p(v_k)\, v_k \,. \tag{1.11}$$

Note that the expected value doesn't have to be one of the possible outcomes. The expected value associated with the roll of a die, for instance, equals 3.5.

To calculate the *rms* deviation from the mean, Δv, we first calculate the squared deviations from the mean, $(v_k - \langle v \rangle)^2$, then we calculate their mean, and finally we take the root:

$$\Delta v = \sqrt{\sum_{k=1}^{K} p(v_k)(v_k - \langle v \rangle)^2} \,. \tag{1.12}$$

The standard deviation of a random variable V with possible values v_k is an important measure—albeit not the only one—of the variability or spread of V.

Problem 1.6. (∗) *Calculate the standard deviation for the sum obtained by rolling two dice.*

Chapter 2

A (very) brief history
of the "old" theory

2.1 Planck

Quantum physics started out as a rather desperate measure to avoid some of the spectacular failures of what we now call "classical physics." The story begins with the discovery by Max Planck, in 1900, of the law that perfectly describes the radiation spectrum of a glowing hot object. (One of the things predicted by classical physics was that you would get blinded by ultraviolet light if you looked at the burner of your stove.) At first it was just a fit to the data—"a fortuitous guess at an interpolation formula," as Planck himself described his radiation law. It was only weeks later that this formula was found to imply the quantization of energy in the emission of electromagnetic radiation, and thus to be irreconcilable with classical physics. According to classical theory, a glowing hot object emits energy *continuously*. Planck's formula implies that it emits energy in *discrete* quantities proportional to the *frequency* ν of the radiation:

$$E = h\nu, \qquad (2.1)$$

where $h = 6.626069 \times 10^{-34}$ Js is the *Planck constant*. Often it is more convenient to use the *reduced Planck constant* $\hbar = h/2\pi$ ("h bar"), which allows us to write

$$E = \hbar\omega, \qquad (2.2)$$

where the *angular frequency* $\omega = 2\pi\nu$ replaces ν.

2.2 Rutherford

In 1911, Ernest Rutherford proposed a model of the atom that was based on experiments conducted by Hans Geiger and Ernest Marsden. Geiger

9

and Marsden had directed a beam of alpha particles (helium nuclei) at a thin gold foil. As expected, most of the alpha particles were deflected by at most a few degrees. Yet a tiny fraction of the particles were deflected through angles much larger than 90 degrees. In Rutherford's own words [Cassidy *et al.* (2002)],

> It was almost as incredible as if you fired a 15-inch shell at a piece of tissue paper and it came back and hit you. On consideration, I realized that this scattering backward must be the result of a single collision, and when I made calculations I saw that it was impossible to get anything of that order of magnitude unless you took a system in which the greater part of the mass of the atom was concentrated in a minute nucleus.

The resulting model, which described the atom as a miniature solar system, with electrons orbiting the nucleus the way planets orbit a star, was however short-lived. Classical electromagnetic theory predicts that an orbiting electron will radiate away its energy and spiral into the nucleus in less than a nanosecond. This was the worst quantitative failure in the history of physics, under-predicting the lifetime of hydrogen by at least forty orders of magnitude. (This figure is based on the experimentally established lower bound on the proton's lifetime.)

2.3 Bohr

In 1913, Niels Bohr postulated that the angular momentum L of an orbiting atomic electron was quantized: its possible values are integral multiples of the reduced Planck constant:

$$L = n\hbar, \quad n = 1, 2, 3 \ldots. \tag{2.3}$$

Observe that angular momentum and Planck's constant are measured in the same units.

Bohr's postulate not only explained the stability of atoms but also accounted for the by then well-established fact that atoms absorb and emit electromagnetic radiation only at specific frequencies. What is more, it enabled Bohr to calculate with remarkable accuracy the spectrum of atomic hydrogen—the particular frequencies at which it absorbs and emits light (visible as well as infrared and ultraviolet).

Apart from his quantization postulate, Bohr's reasoning at the time remained completely classical. Let us assume with Bohr that the electron's

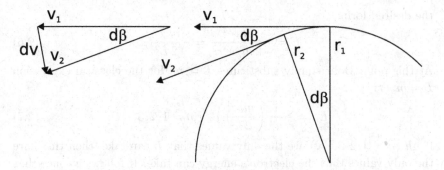

Fig. 2.1   Calculating the acceleration of an orbiting electron.

orbit is a circle of radius r. The electron's speed is then given by $v = r\,d\beta/dt$, where $d\beta$ is the small angle traversed during a short time dt, while the magnitude a of the electron's acceleration is the magnitude dv of the vector difference $\mathbf{v}_2 - \mathbf{v}_1$ divided by dt.[1] This equals $a = v\,d\beta/dt$, as we gather from Fig. 2.1. Eliminating $d\beta/dt$ by using $v = r\,d\beta/dt$, we arrive at $a = v^2/r$.

We want to calculate the electron's total energy as it orbits the nucleus (a proton). In Gaussian units, the magnitude of the Coulomb force exerted on the electron by the proton takes the particularly simple form $F = e^2/r^2$, where e is the absolute value of both the electron's and the proton's charge. Since $F = ma = mv^2/r$, we have that $mv^2 = e^2/r$. This gives us the electron's *kinetic* energy,

$$E_K = \frac{m_e v^2}{2} = \frac{e^2}{2r}, \tag{2.4}$$

where m_e is the electron's mass.

By convention, the electron's *potential* energy is 0 at $r = \infty$. Its potential energy at the distance r from the nucleus is therefore minus the work done by moving it from r to infinity,

$$E_P = -\int_r^\infty F\,dr = -\int_r^\infty \frac{e^2}{(r')^2}\,dr' = -\frac{e^2}{r}. \tag{2.5}$$

(You will do the integral in the next chapter.) So the electron's total energy is $E = E_K + E_P = -e^2/2r$.

Our next order of business is to express E as a function of L rather than r. Classically, $L = m_e vr$. Equation (2.4) allows us to massage E into

---

[1]To be precise, this holds in the limit in which dt, and hence $d\beta$ and dv, go to 0. See the next chapter for a brief introduction to vectors, differential quotients, and such.

the desired form:

$$E = -\frac{m_e e^4}{2\,m_e^2 v^2 r^2} = -\frac{m_e e^4}{2\,L^2}\,. \tag{2.6}$$

At this point Bohr simply substitutes $L = n\hbar$ for the classical expression $L = m_e vr$:

$$E_n = -\frac{1}{n^2}\left(\frac{m_e e^4}{2\,\hbar^2}\right), \quad n = 1, 2, 3, \ldots \tag{2.7}$$

If $n\hbar$ $(n = 1, 2, 3, \ldots)$ are the only values that L can take, then these are the only values that the electron's energy can take. It follows at once that a hydrogen atom can emit or absorb energy only by amounts equal to the differences

$$\Delta E_{nm} = E_n - E_m = \left(\frac{1}{m^2} - \frac{1}{n^2}\right) \text{Ry}\,, \tag{2.8}$$

where the *Rydberg* (Ry) is an energy unit equal to $m_e e^4/2\hbar^2 = 13.605691\,\text{eV}$. It is also the ionization energy $\Delta E_{\infty 1}$ of atomic hydrogen in its ground state.

Considering the variety of wrong classical assumptions that went into the derivation of Eq. (2.8), it is remarkable that the frequencies predicted by Bohr via $\nu_{nm} = E_{nm}/h$ were in excellent agreement with the experimentally known frequencies at which atomic hydrogen emits and absorbs light.

2.4   de Broglie

In 1923, ten years after Bohr postulated that L comes in integral multiples of $\hbar$, someone finally hit on an explanation *why* angular momentum was quantized. In 1905, Albert Einstein had argued that electromagnetic radiation itself was quantized—not merely its emission and absorption, as Planck had held. Planck's radiation formula had implied a relation between a particle property and a wave property for the quanta of electromagnetic radiation we now call *photons*: $E = h\nu$. Einstein's explanation of the photoelectric effect established another such relation:

$$p = h/\lambda\,, \tag{2.9}$$

where p is the photon's momentum and λ is its wavelength. But if electromagnetic waves have particle properties, Louis de Broglie reasoned, why cannot electrons have wave properties?

Imagine that the electron in a hydrogen atom is a standing wave on a circle (Fig. 2.2) rather than a corpuscle moving in a circle. (The crests,

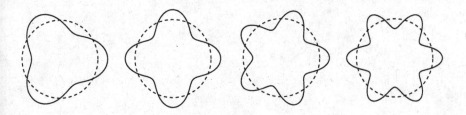

Fig. 2.2   Standing waves on a circle for $n = 3, 4, 5, 6$.

troughs, and nodes of a standing wave are stationary—they stay put.) Such a wave has to satisfy the condition

$$2\pi r = n\lambda, \quad n = 1, 2, 3, \ldots, \tag{2.10}$$

i.e., the circle's circumference $2\pi r$ must be an integral multiple of λ. Using $p = h/\lambda$ to eliminate λ from Eq. (2.10) yields $pr = n\hbar$. But $pr = mvr$ is just the angular momentum L of a classical electron moving in a circle of radius r. In this way de Broglie arrived at the quantization condition $L = n\hbar$, which Bohr had simply postulated.

Chapter 3

Mathematical interlude

3.1 Vectors

A *vector* is a quantity that has both a magnitude and a direction—for present purposes a direction in "ordinary" 3-dimensional space. Such a quantity can be represented by an arrow.

The sum of two vectors can be defined via the parallelogram rule: (i) move the arrows (without changing their magnitudes or directions) so that their tails coincide, (ii) duplicate the arrows, (iii) move the duplicates (again without changing magnitudes or directions) so that (a) their tips coincide and (b) the four arrows form a parallelogram. The resultant vector extends from the tails of the original arrows to the tips of their duplicates.

If we introduce a coordinate system with three mutually perpendicular axes, we can characterize a vector $\mathbf{a}$ by its components (a_x, a_y, a_z) (Fig. 3.1).

Problem 3.1. ($*$) *The sum* $\mathbf{c} = \mathbf{a} + \mathbf{b}$ *of two vectors has the components* $(c_x, c_y, c_z) = (a_x + b_x, a_y + b_y, a_z + b_z)$.

The *dot product* of two vectors $\mathbf{a}, \mathbf{b}$ is the number

$$\mathbf{a} \cdot \mathbf{b} \stackrel{\text{Def}}{=} a_x b_x + a_y b_y + a_z b_z \,. \tag{3.1}$$

We need to check that this definition is independent of the (rectangular) coordinate system to which the vector components on the right-hand side refer. To this end we calculate

$$
\begin{aligned}
(\mathbf{a} + \mathbf{b}) \cdot (\mathbf{a} + \mathbf{b}) &= (a_x + b_x)^2 + (a_y + b_y)^2 + (a_z + b_z)^2 \\
&= a_x^2 + a_y^2 + a_z^2 + b_x^2 + b_y^2 + b_z^2 + 2\,(a_x b_x + a_y b_y + a_z b_z) \\
&= \mathbf{a} \cdot \mathbf{a} + \mathbf{b} \cdot \mathbf{b} + 2\,\mathbf{a} \cdot \mathbf{b} \,. \tag{3.2}
\end{aligned}
$$

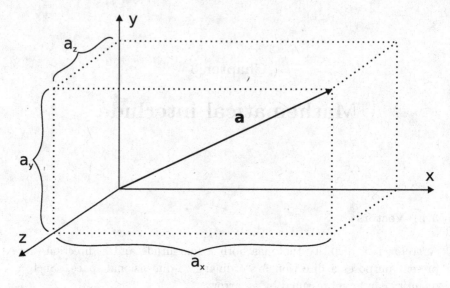

Fig. 3.1 The components of a vector.

According to Pythagoras, the magnitude a of a vector $\mathbf{a}$ equals $\sqrt{a_x^2 + a_y^2 + a_z^2}$. Because the left-hand side and the first two terms on the right-hand side of Eq. (3.2) are the squared magnitudes of vectors, they do not change under a coordinate transformation that preserves the magnitudes of all vectors. Hence the third term on the right-hand side does not change under such a transformation, and neither therefore does the product $\mathbf{a} \cdot \mathbf{b}$. But the coordinate transformations that preserve the magnitudes of vectors also preserve the angles between vectors. In particular, they turn a system of rectangular coordinates into another system of rectangular coordinates. Thus while the individual components on the right-hand side of Eq. (3.2) generally change under such a transformation, the dot product $\mathbf{a} \cdot \mathbf{b}$ does not.

By the term *scalar* we mean a number that is invariant under transformations of some kind or other. Since the dot product is invariant under translations and rotations of the coordinate axes—the transformations that preserve magnitudes and angles—it is also known as *scalar product*.

Problem 3.2. (∗) $\mathbf{a} \cdot \mathbf{b} = ab \cos\theta$, *where θ is the angle between* $\mathbf{a}$ *and* $\mathbf{b}$.

Another useful definition (albeit only in a 3-dimensional space) is the *cross product* of two vectors. If $\hat{\mathbf{x}}, \hat{\mathbf{y}}, \hat{\mathbf{z}}$ are unit vectors parallel to the coordinate

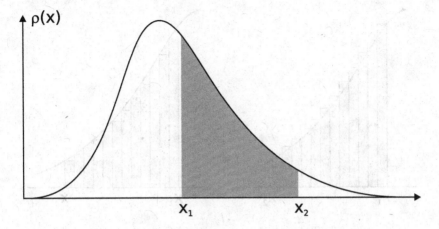

Fig. 3.2 The area corresponding to a definite integral.

axes, this is given by

$$\mathbf{a} \times \mathbf{b} \stackrel{\text{Def}}{=} (a_y b_z - a_z b_y)\,\hat{\mathbf{x}} + (a_z b_x - a_x b_z)\,\hat{\mathbf{y}} + (a_x b_y - a_y b_x)\,\hat{\mathbf{z}}. \qquad (3.3)$$

Problem 3.3. *The cross product is antisymmetric:* $\mathbf{a} \times \mathbf{b} = -\mathbf{b} \times \mathbf{a}$.

Problem 3.4. (∗) $\mathbf{a} \times \mathbf{b}$ *is perpendicular to both* $\mathbf{a}$ *and* $\mathbf{b}$.

Problem 3.5. $\hat{\mathbf{x}} \times \hat{\mathbf{y}} = \hat{\mathbf{z}}$, $\hat{\mathbf{y}} \times \hat{\mathbf{z}} = \hat{\mathbf{x}}$, $\hat{\mathbf{z}} \times \hat{\mathbf{x}} = \hat{\mathbf{y}}$.

By convention, the direction of $\mathbf{a} \times \mathbf{b}$ is given by the right-hand rule: if the first (index) and the second (middle) finger of your right hand point in the direction of $\mathbf{a}$ and $\mathbf{b}$, respectively, then your right thumb (pointing in a direction perpendicular to both $\mathbf{a}$ and $\mathbf{b}$) indicates the direction of $\mathbf{a} \times \mathbf{b}$.

Problem 3.6. (∗) *The magnitude of* $\mathbf{a} \times \mathbf{b}$ *equals* $ab\sin\theta$, *the area of the parallelogram spanned by* $\mathbf{a}$ *and* $\mathbf{b}$.

3.2 Definite integrals

We frequently have to deal with probabilities that are assigned to intervals of a continuous variable x (like the interval $[x_1, x_2]$ in Fig. 3.2). Such probabilities are calculated with the help of a *probability density function* $\rho(x)$, which is defined so that the probability with which x is found to

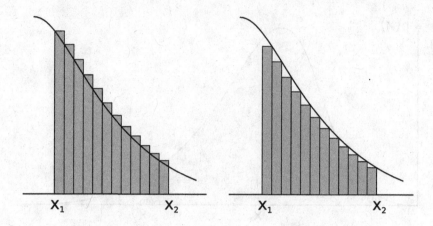

Fig. 3.3 Two approximations to the definite integral (3.4).

lie in the interval $[x_1, x_2]$ is given by the shaded area in Fig. 3.2. The mathematical tool for calculating this area is the *(definite) integral*

$$A = \int_{x_1}^{x_2} \rho(x)\,dx. \tag{3.4}$$

To define this integral, we overlay the shaded area of Fig. 3.2 with N rectangles of width $\Delta x = (x_2 - x_1)/N$ in either of the ways shown in Fig. 3.3. The sum of the rectangles in the left half of this figure,

$$A_+ = \sum_{k=0}^{N-1} \rho(x + k\,\Delta x)\,\Delta x, \tag{3.5}$$

is larger than the wanted area A, while the sum of the rectangles in the right half,

$$A_- = \sum_{k=1}^{N} \rho(x + k\,\Delta x)\,\Delta x, \tag{3.6}$$

is smaller. It is clear, though, that the differences $A_+ - A$ and $A - A_-$ decrease as the number of rectangles increases. The integral (3.4) is defined as the limit of either sum:

$$\lim_{N \to \infty} \sum_{k=1}^{N} \rho(x + k\,\Delta x)\,\Delta x = \int_{x_1}^{x_2} \rho(x)\,dx = \lim_{N \to \infty} \sum_{k=0}^{N-1} \rho(x + k\,\Delta x)\,\Delta x.$$

Another frequently used expression is the integral $\int_{-\infty}^{+\infty} \rho(x)\,dx$, which is defined as the limit

$$\lim_{a \to \infty} \int_{-a}^{+a} \rho(x)\,dx. \tag{3.7}$$

One often has to integrate functions of more than one variable. Take the integral

$$\int_R f(x, y, z) \, d^3r \,. \tag{3.8}$$

R is a region of 3-space, and $d^3r = dx \, dy \, dz$ is the volume of an infinitely small rectangular cuboid with sides dx, dy, dz. Instead of summing over infinitely many infinitely small intervals lying inside a finite interval, one now sums over infinitely many infinitely small rectangular cuboids lying inside a finite region R. (For more on infinitely many infinitely small things see the next section.)

3.3 Derivatives

A *function* $f(x)$ is a machine that has an input and an output. Insert a number x and out pops the number $f(x)$. [*Warning*: sometimes $f(x)$ denotes the machine itself rather than the number obtained after inserting a particular x.] We shall mostly be dealing with functions that are well-behaved. Saying that a function $f(x)$ is *well-behaved* is the same as saying that we can draw its graph without lifting up the pencil, and we can do the same with the graphs of its derivatives.

The (first) *derivative* of $f(x)$ is a machine $f'(x)$ that works like this: insert a number x, and out pops the slope of (the graph of) $f(x)$ at x. What we mean by the *slope* of $f(x)$ at a particular point $x = a$ is the slope of the *tangent* $t(x)$ on $f(x)$ at a.

Take a look at Fig. 3.4. The curve in each of the three diagrams is (the graph of) $f(x)$. The slope of the straight line $s(x)$ that intersects $f(x)$ at two points in the upper diagrams is given by the *difference quotient*

$$\frac{\Delta s}{\Delta x} = \frac{s(x + \Delta x) - s(x)}{\Delta x} \,. \tag{3.9}$$

This tells us how much $s(x)$ increases as x increases by Δx. The lower diagram shows the tangent $t(x)$ on the function $f(x)$ for a particular x.

Now consider the small black disk at the intersection of the functions $f(x)$ and $s(x)$ at $x + \Delta x$ in the upper left diagram. Think of it as a bead sliding along $f(x)$ towards the left. As it does so, the slope of $s(x)$ increases (compare the upper two diagrams). In the limit in which this bead occupies the same place as the bead sitting at x, $s(x)$ coincides with $t(x)$, as one gleans from the lower diagram. In other words, as Δx tends to 0, the

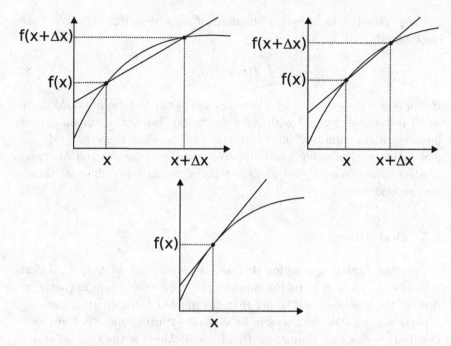

Fig. 3.4 Definition of the slope of a function $f(x)$ at x.

difference quotient (3.9) tends to the *differential quotient*

$$\frac{df}{dx} \stackrel{\text{Def}}{=} \lim_{\Delta x \to 0} \frac{\Delta f}{\Delta x}, \qquad (3.10)$$

which is the same as $f'(x)$. The *differentials* dx and df are *infinitesimal* ("infinitely small") quantities. This sounds highly mysterious until one realizes that every expression containing such quantities is to be understood as the limit in which these tend to 0, one (here, dx) independently, the others (here, df) dependently.

To *differentiate* a function $f(x)$ is to obtain its first derivative $f'(x)$. By differentiating $f'(x)$, we obtain the second derivative $f''(x)$ of $f(x)$, for which we can also write $d^2 f/dx^2$. To make sense of the last expression, think of d/dx as an operator. Like a function, an *operator* has an input and an output, but unlike a function, it accepts a function as input. Insert $f(x)$ into d/dx and get the function df/dx. Insert the output of d/dx into another operator d/dx and get the function $(d/dx)(d/dx)f(x) \stackrel{\text{Def}}{=} (d^2/dx^2)f(x) = d^2 f/dx^2$.

By differentiating the second derivative we obtain the third, and so on.

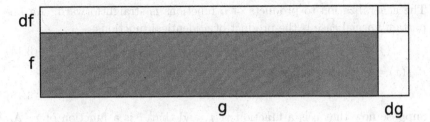

Fig. 3.5   Illustration of the product rule.

Problem 3.7. *Find the slope of the straight line* $f(x) = ax + b$.

Problem 3.8. (∗) *Calculate* $f'(x)$ *for* $f(x) = 2x^2 - 3x + 4$.

Problem 3.9. (∗) *What does* $f''(x)$—*the slope of the slope of* $f(x)$—*tell us about the graph of* $f(x)$?

By definition, $(f + g)(x) = f(x) + g(x)$.

Problem 3.10. *If a is a number and f and g are functions of x, then*

$$\frac{d(af)}{dx} = a\frac{df}{dx} \quad and \quad \frac{d(f + g)}{dx} = \frac{df}{dx} + \frac{dg}{dx}.$$

A slightly more difficult task is to differentiate the product $h(x) = f(x)\,g(x)$. Think of f and g as the vertical and horizontal sides of a rectangle of area h. As x increases by Δx, the product fg increases by the sum of the areas of the three white rectangles in Fig. 3.5:

$$\Delta h = f(\Delta g) + (\Delta f)g + (\Delta f)(\Delta g). \tag{3.11}$$

Hence

$$\frac{\Delta h}{\Delta x} = f\frac{\Delta g}{\Delta x} + \frac{\Delta f}{\Delta x}g + \frac{\Delta f\,\Delta g}{\Delta x}. \tag{3.12}$$

If we now let Δx go to 0, the first two terms on the right-hand side tend to $fg' + f'g$. What about the third term? Since it is the product of an expression (either $\Delta g/\Delta x$ or $\Delta f/\Delta x$) that tends to a finite number and an expression (either Δf or Δg) that tends to 0, it tends to 0. The bottom line:

$$h' = (f\,g)' = f\,g' + f'\,g. \tag{3.13}$$

Problem 3.11. (∗) $(f\,g\,h)' = f\,g\,h' + f\,g'\,h + f'\,g\,h$.

The generalization to products of n functions is straightforward. An important special case is the product of n identical functions:

$$(f^n)' = f^{n-1} f' + f^{n-2} f' f + \cdots + f' f^{n-1} = n f^{n-1} f'. \tag{3.14}$$

If $f(x) = x$, this boils down to

$$(x^n)' = n x^{n-1}. \tag{3.15}$$

Suppose now that g is a function of f, and that f is a function of x. An increase in x by Δx will cause an increase in f by $\Delta f \approx (df/dx)\Delta x$, and this will cause an increase in g by $\Delta g \approx (dg/df)\Delta f$ (the symbol $\approx$ means "is approximately equal to"). Thus

$$\frac{\Delta g}{\Delta x} \approx \frac{dg}{df}\frac{df}{dx}. \tag{3.16}$$

In the limit $\Delta x \to 0$, "approximately equal" becomes "equal," and Eq. (3.16) becomes the *chain rule*

$$\frac{dg}{dx} = \frac{dg}{df}\frac{df}{dx}. \tag{3.17}$$

Problem 3.12. *We have proved Eq. (3.15) for integers $n \geq 2$. Check that it also holds for $n = 0$ and $n = 1$.*

Problem 3.13. (∗) *Equation (3.15) also holds for negative integers n.*

Problem 3.14. (∗) *Equation (3.15) also holds for $n = 1/m$, where m is a natural number.*

Problem 3.15. *Use the chain rule (3.17) to show that if Eq. (3.15) holds for $n = a$ and $n = b$, then it also holds for $n = ab$.*

It follows from what you have just shown that Eq. (3.15) holds for all *rational numbers* n. Moreover, since every real number is the limit of a *sequence* of rational numbers, we can make sure that Eq. (3.15) holds for all *real numbers*, by defining it as the limit of some sequence in case n is an irrational number.

We often use functions with more than one input slot. The output of $f(t, x, y, z)$, for example, depends on the time coordinate t as well as the spatial coordinates x, y, z. If we choose a fixed set of values x, y, z, we obtain a function $f_{xyz}(t)$ of t alone. The *partial derivative* of $f(t, x, y, z)$ with respect to t is the derivative of $f_{xyz}(t)$, for which we write $\partial f/\partial t$ (usually without explicitly indicating that this function depends on the chosen set of values x, y, z). The partial derivatives of $f(t, x, y, z)$ with respect to the other variables are defined analogously.

3.4 Taylor series

A well-behaved function can be expanded into a power series. This means that for all non-negative integers $k = 0, 1, 2, \ldots$ there are real numbers a_k such that

$$f(x) = \sum_{k=0}^{\infty} a_k x^k = a_0 + a_1 x + a_2 x^2 + a_3 x^3 + a_4 x^4 + \cdots . \qquad (3.18)$$

Let's calculate the first four derivatives using (3.15):

$$f'(x) = a_1 + 2 a_2 x + 3 a_3 x^2 + 4 a_4 x^3 + 5 a_5 x^4 + \cdots ,$$
$$f''(x) = 2 a_2 + 2 \cdot 3 a_3 x + 3 \cdot 4 a_4 x^2 + 4 \cdot 5 a_5 x^3 + \cdots ,$$
$$f'''(x) = 2 \cdot 3 a_3 + 2 \cdot 3 \cdot 4 a_4 x + 3 \cdot 4 \cdot 5 a_5 x^2 + \cdots ,$$
$$f''''(x) = 2 \cdot 3 \cdot 4 a_4 + 2 \cdot 3 \cdot 4 \cdot 5 a_5 x + \cdots .$$

Setting x equal to zero, we obtain the following values:

$$f(0) = a_0 , \quad f'(0) = a_1 , \quad f''(0) = 2 a_2 ,$$
$$f'''(0) = 2 \times 3 a_3 , \quad f''''(0) = 2 \times 3 \times 4 a_4 .$$

Since we don't want to go on adding primes ($'$), we will write $f^{(n)}(x)$ for the n-th derivative of $f(x)$. If we also write $f^{(0)}(x)$ for $f(x)$, we have that $f^{(k)}(0)$ equals $k! \, a_k$, where the *factorial* $k!$ is defined as equal to 1 for $k = 0$ and $k = 1$, and as the product of all natural numbers $n \le k$ for $k > 1$. Expressing the *coefficients* a_k in terms of the derivatives of $f(x)$ for $x = 0$, we arrive at the following power series—also known as the *Taylor series*—for $f(x)$:

$$f(x) = \sum_{k=0}^{\infty} \frac{f^{(k)}(0)}{k!} x^k . \qquad (3.19)$$

A remarkable result: if you know the value of a well-behaved function $f(x)$ and the values of *all* of its derivatives for a single value of x (in this case $x = 0$, but there is a similar series for any value of x), then you know $f(x)$ for *all* values of x.

3.5 Exponential function

We define the function $\exp(x)$ by requiring that $\exp'(x) = \exp(x)$ and $\exp(0) = 1$. In other words, the value of this function is everywhere equal to the slope of its graph, which intersects the vertical axis at the value 1.

Problem 3.16. *Sketch the graph of* $\exp(x)$ *using this information alone.*

Problem 3.17. *All derivatives of* $\exp(x)$ *are equal to* $\exp(x)$.

Thus $\exp^{(k)}(0) = 1$ for all k, whence a particularly simple Taylor series results:

$$\exp(x) = \sum_{k=0}^{\infty} \frac{x^k}{k!} = 1 + x + \frac{x^2}{2} + \frac{x^3}{6} + \frac{x^4}{24} + \cdots. \tag{3.20}$$

Problem 3.18. (∗) $\exp(x)$ *satisfies*

$$f(a)\,f(b) = f(a+b). \tag{3.21}$$

It can be shown that every function satisfying Eq. (3.21) has the form $f(x) = a^x$. This means that there is a number e such that $\exp(x) = e^x$— hence the name "exponential function."

Problem 3.19. (∗) *Calculate* e.

Problem 3.20. $d(e^{ax})/dx = a\,e^{ax}$.

The *natural logarithm* $\ln x$ is the inverse of e^x, that is, $e^{\ln x} = \ln(e^x) = x$.

Problem 3.21. $\ln a + \ln b = \ln(ab)$.

Problem 3.22. (∗)

$$\frac{d\ln f(x)}{dx} = \frac{1}{f(x)}\frac{df}{dx}. \tag{3.22}$$

3.6   Sine and cosine

We define the function $\cos(x)$ by requiring that $\cos''(x) = -\cos(x)$, $\cos(0) = 1$, and $\cos'(0) = 0$.

Problem 3.23. (∗) *Sketch the graph of* $\cos(x)$, *making use of this information alone.*

Problem 3.24. *For* $n \geq 0$: $\cos^{(n+2)}(x) = -\cos^{(n)}(x)$.

Problem 3.25.

$$\cos^{(k)}(0) = \begin{cases} +1 \text{ for } k = 0, 4, 8, 12, \ldots \\ -1 \text{ for } k = 2, 6, 10, 14, \ldots \\ 0 \quad \text{ for odd } k \end{cases}$$

We thus arrive at the following Taylor series:

$$\cos(x) = 1 - \frac{x^2}{2!} + \frac{x^4}{4!} - \frac{x^6}{6!} + \cdots . \tag{3.23}$$

The function $\sin(x)$ is defined by requiring that $\sin''(x) = -\sin(x)$, $\sin(0) = 0$, and $\sin'(0) = 1$. This leads to the Taylor series

$$\sin(x) = x - \frac{x^3}{3!} + \frac{x^5}{5!} - \frac{x^7}{7!} + \cdots . \tag{3.24}$$

3.7 Integrals

In Sec. 3.2 we defined the definite integral as a limit. How do we calculate this limit? The answer is elementary if we know a function $F(x)$ of which $f(x)$ is the first derivative, $f = dF/dx$, for we can then substitute dF for $f\,dx$:

$$\int_a^b f(x)\,dx = \int_a^b dF(x) . \tag{3.25}$$

On the face of it, we are still adding infinitely many infinitely small quantities, but look what this amounts to:

$$\int_a^b dF(x) = [F(a + dx) - F(a)]$$
$$+ [F(a + 2\,dx) - F(a + dx)]$$
$$+ [F(a + 3\,dx) - F(a + 2\,dx)]$$
$$| \cdots$$
$$+ [F(b - 2\,dx) - F(b - 3\,dx)]$$
$$+ [F(b - dx) - F(b - 2\,dx)]$$
$$+ [F(b) - F(b - dx)] .$$

After all cancellations are done, we are left with $\int_a^b dF(x) = F(b) - F(a)$.

If $f(x)$ is the derivative of $F(x)$, $F(x)$ is known as an *integral* or *antiderivative* of $f(x)$—*an* integral rather than *the* integral because if $F(x)$ is an integral of $f(x)$ and c is a constant, then $F(x) + c$ is another integral of $f(x)$. To distinguish integrals from definite integrals, we also refer to them as *indefinite* integrals.

Problem 3.26. (∗) *Calculate $\int_1^2 x^2\,dx$.*

Problem 3.27. (∗)

$$\int_r^\infty \frac{1}{(r')^2}\, dr' = \frac{1}{r}\,.$$

If we don't know an antiderivative of $f(x)$, calculating the integral $\int_a^b dx\, f(x)$ is much harder. Let's do the *Gaussian integral*,

$$I = \int_{-\infty}^{+\infty} dx\, e^{-x^2/2}, \qquad (3.26)$$

as a case in point. For this integral someone has discovered the following trick. (The trouble is that different integrals usually require different tricks.) Start with the square of I:

$$I^2 = \int_{-\infty}^{+\infty} dx\, e^{-x^2/2} \int_{-\infty}^{+\infty} dy\, e^{-y^2/2} = \int_{-\infty}^{+\infty}\int_{-\infty}^{+\infty} dx\, dy\, e^{-(x^2+y^2)/2}.$$

This is an integral over the x–y plane. We are again adding infinitely many infinitely small quantities, in this case rectangles of area $dx\, dy$, each multiplied by the value that the *integrand* $e^{-(x^2+y^2)/2}$ takes somewhere inside it.

Now let's reduce this double integral to a single one by switching to *polar coordinates*.[1] For $x^2 + y^2$ we substitute r^2, and instead of summing contributions from infinitesimal rectangles we sum contributions from infinitesimal annuli of area $2\pi r\, dr$.[2] Finally, to cover the entire plane, we let r range from 0 to ∞:

$$I^2 = 2\pi \int_0^{+\infty} dr\, r\, e^{-r^2/2}.$$

Now we use $d\,r^2/dr = 2r$ to replace $dr\, r$ by $d(r^2/2)$, and we substitute a new integration variable w for $r^2/2$:

$$I^2 = 2\pi \int_0^{+\infty} d\left(r^2/2\right) e^{-r^2/2} = 2\pi \int_0^{+\infty} dw\, e^{-w}.$$

We are almost done, since the antiderivative of e^{-w} is known. It is $-e^{-w}$. Hence

$$\int_0^{+\infty} dw\, e^{-w} = (-e^{-\infty}) - (-e^0) = 0 + 1 = 1\,.$$

---

[1] For the definition of polar coordinates in three dimensions, see Fig. 11.2.

[2] The area of an annulus with inner and outer radii r and $r + dr$ is given by $\pi(r + dr)^2 - \pi r^2 = 2\pi r\, dr + \pi\, dr^2$. But since the limit $dr \to 0$ is implied, all but the lowest order of dr can be ignored. (Recall our derivation of the product rule 3.13.)

So $I^2 = 2\pi$ and

$$I = \int_{-\infty}^{+\infty} dx\, e^{-x^2/2} = \sqrt{2\pi}\,. \tag{3.27}$$

Believe it or not, a significant fraction of the literature in theoretical physics concerns variations and elaborations of this integral.

One such variation can be obtained by substituting $\sqrt{a}\,x$ for x:

$$\int_{-\infty}^{+\infty} dx\, e^{-ax^2/2} = \sqrt{\frac{2\pi}{a}}\,. \tag{3.28}$$

Another variation can be obtained by treating both sides of this equation as functions of a and differentiating them with respect to a. The result is

$$\int_{-\infty}^{+\infty} dx\, e^{-ax^2/2} x^2 = \sqrt{\frac{2\pi}{a^3}}\,. \tag{3.29}$$

Problem 3.28. *Prove the last two equations.*

One method that sometimes helps evaluating an integral is known as *integration by parts*. Integrating the product rule (3.13) yields

$$\int_a^b dx(fg)' = \int_a^b dx fg' + \int_a^b dx f'g\,. \tag{3.30}$$

This allows us to write

$$\int_a^b dx fg' = (fg)(b) - (fg)(a) - \int_a^b dx f'g\,. \tag{3.31}$$

3.8 Complex numbers

"God created the natural numbers, all the rest is the work of man," the mathematician Leopold Kronecker is reported to have said. By subtracting *natural numbers* from natural numbers, we can create *integers* that are not natural numbers. By dividing integers by integers we can create *rational numbers* that are not integers. By taking the limits of sequences of rational numbers—or by doing something more specific, like taking the square roots of positive integers—we can create *real numbers* that are not rational. And by taking the roots of polynomials we can create *complex numbers* that are not real.

A *polynomial* $p(x)$ is like a power series except that it only contains a finite number of terms. The *roots* of a polynomial $p(x)$ are the values of x for which $p(x) = 0$. Take the polynomial $p(x) = 1 + x^2$. What are its roots? You might be tempted to say that they do not exist. If so, what stops us from inventing them? It's as easy as saying: "Let i be equal to the positive square root of -1!" All you can say is that the roots $+i$ and $-i$ of $p(x)$ are not real numbers. They are imaginary numbers.

Do not be misled by the conventional labels "real" and "imaginary." No number is real in the sense in which, say, apples are real. Both real numbers and complex numbers are creations of the human mind. The real numbers are no less imaginary (in the ordinary sense of "imaginary") than the imaginary numbers. All you can say is that you have been using natural numbers (for counting), rational numbers (for accounting), and real numbers (for measuring), whereas you haven't yet found a use for complex numbers. But this is going to change. Quantum mechanics *requires* the use of complex numbers.

An *imaginary number* is a real number multiplied by $i = +\sqrt{-1}$. Every *complex number* z is the sum of a real number a (the *real part* of z) and an imaginary number ib. Somewhat confusingly, the *imaginary part* of z is the *real* number b.

Because real numbers may be visualized as points in a line, the set of real numbers is sometimes called the *real line*. Because complex numbers may be visualized as points in a plane, the set of complex numbers is often referred to as the *complex plane*. This plane contains two axes, one horizontal (the *real axis* containing the real numbers) and one vertical (the *imaginary axis* containing the imaginary numbers).

Figure 3.6 illustrates the addition of two complex numbers:

$$z_1 + z_2 = (a_1 + ib_1) + (a_2 + ib_2) = (a_1 + a_2) + i(b_1 + b_2). \qquad (3.32)$$

We often think of complex numbers as arrows which, like the vectors we considered in Sec. 3.1, have a magnitude and a direction, but no particular location. It is readily seen that adding two complex numbers, considered as arrows, works just like adding vectors in a plane.

To be able to multiply complex numbers, all you need to know is that $i^2 = -1$:

$$z_1 z_2 = (a_1 + ib_1)(a_2 + ib_2) = (a_1 a_2 - b_1 b_2) + i(a_1 b_2 + b_1 a_2). \qquad (3.33)$$

A useful definition is the *complex conjugate* $z^* = a - ib$ of $z = a + ib$. Among other things, it allows us to write $z^* z$ for the square of the absolute value $|z|$ of z—the *absolute square* $|z|^2$, for short.

Problem 3.29. $zz^* = a^2 + b^2$.

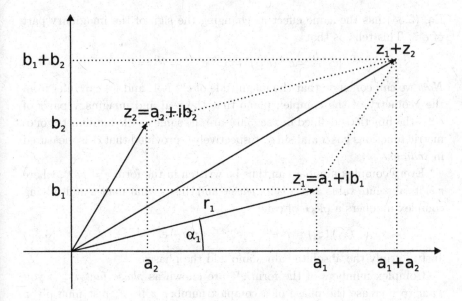

Fig. 3.6   Adding complex numbers.

There is an easier way of multiplying complex numbers. To see how it works, let us first redefine the exponential function in terms of its Taylor series. This allows us to write down the Taylor series for e^{ix}:

$$\sum_{k=0}^{\infty} \frac{(ix)^k}{k!} = 1 + ix + \frac{(ix)^2}{2!} + \frac{(ix)^3}{3!} + \frac{(ix)^4}{4!} + \frac{(ix)^5}{5!} + \frac{(ix)^6}{6!} + \frac{(ix)^7}{7!} + \cdots .$$

Problem 3.30. $i^3 = -i$ *and* $i^4 = 1$.

Problem 3.31. (∗) *The real and imaginary parts of* $e^{i\alpha}$ *are* $\cos\alpha$ *and* $\sin\alpha$, *respectively*.

The result is *Euler's formula*:

$$e^{i\alpha} = \cos\alpha + i\sin\alpha . \tag{3.34}$$

What is the magnitude of $e^{i\alpha}$? Its absolute square works out at

$$e^{i\alpha}(e^{i\alpha})^* = (\cos\alpha + i\sin\alpha)(\cos\alpha - i\sin\alpha) = \cos^2\alpha + \sin^2\alpha .$$

Remembering the trigonometric relation $\cos^2\alpha + \sin^2\alpha = 1$, you may be tempted to conclude that $|e^{i\alpha}| = 1$. But we haven't yet shown that the functions $\cos(x)$ and $\sin(x)$ defined in Sec. 3.6 are the same as their trigonometric namesakes! To do so, we observe that changing the sign of α in

Eq. (3.34) has the same effect as changing the sign of the imaginary part of $e^{i\alpha}$. This tells us that

$$|e^{i\alpha}|^2 = e^{i\alpha}(e^{i\alpha})^* = e^{i\alpha}\,e^{-i\alpha} = e^0 = 1\,.$$

Now we can conclude that the magnitude of $e^{i\alpha}$ is 1, and we can infer from the geometry of the complex plane that the real and imaginary parts of $e^{i\alpha}$—the functions defined in Sec. 3.6—are the same as the familiar trigonometric functions $\cos\alpha$ and $\sin\alpha$, respectively—provided that α is measured in *radians*.[3]

Every complex number can thus be written in the form $z = re^{i\alpha}$, where r is its absolute value and α is its *phase*, and this is what makes multiplying complex numbers a piece of cake:

$$(z_1)(z_2) = r_1 e^{i\alpha_1}\, r_2 e^{i\alpha_2} = (r_1 r_2)\,e^{i(\alpha_1 + \alpha_2)}. \tag{3.35}$$

Just multiply the absolute values and add the phases.

Complex numbers of the form e^{ix} are known as *phase factors*. If you want to increase the phase of a complex number z by β, just multiply z by $e^{i\beta}$.

Problem 3.32. (∗) *Find the real and imaginary parts of $e^{i\pi/4}$.*

Problem 3.33. (∗) *Find* all *roots of $x^2 - i = 0$ and $x^3 + 1 = 0$.*

Problem 3.34.

$$\cos x = \frac{e^{ix} + e^{-ix}}{2}\,, \qquad \sin x = \frac{e^{ix} - e^{-ix}}{2i}\,.$$

Problem 3.35. (∗) *Arguably the five most important numbers in mathematics are $0, 1, i, \pi$, and e. Write down a correct equation that contains each of them just once.*

---

[3]Measured in radians, the angle subtended at the center of a circle of radius 1 by an arc of length L equals L. $360°$ thus corresponds to 2π.

Chapter 4

A (very) brief history
of the "new" theory

4.1 Schrödinger

If the electron is a standing wave, why should it be confined to a circle?
After the insight that particles can behave like waves, which de Broglie
gained ten years after Bohr postulated the quantization of angular momentum,
it took less than three years for the mature ("new") quantum theory
to be formulated, not once but twice in different mathematical attire, by
Werner Heisenberg in 1925 and by Erwin Schrödinger in 1926.

If we imagine the electron as a standing wave *in three dimensions*, we
have almost all it takes to arrive at the equation that is at the heart of the
new theory. To keep things simple, let us however start with one spatial dimension.
The simplest mathematical description of a wave of *amplitude* A,
wavenumber $k = 2\pi/\lambda$, and *angular frequency* $\omega = 2\pi/T = 2\pi\nu$ is the
function (Fig. 4.1)

$$\psi(x,t) = A\,e^{i(kx-\omega t)}. \tag{4.1}$$

Expressing the phase $\phi(x,t) = kx - \omega t$ in terms of the electron's energy
$E = \hbar\omega$ (Eq. 2.2) and momentum $p = h/\lambda = \hbar k$ (Eq. 2.9),

$$\psi(x,t) = A\,e^{i(px-Et)/\hbar}, \tag{4.2}$$

and taking the first partial derivative with respect to t and the second
partial derivative with respect to x, we obtain

$$\frac{\partial\psi}{\partial t} = -\frac{i}{\hbar}\,E\psi\,, \qquad \frac{\partial^2\psi}{\partial x^2} = -\frac{1}{\hbar^2}\,p^2\,\psi\,. \tag{4.3}$$

In the classical and non-relativistic theory, the energy E and the momentum p of a freely moving particle are related via $E = p^2/2m$. (This can
be made to look more familiar by substituting mv for p. We shall discover

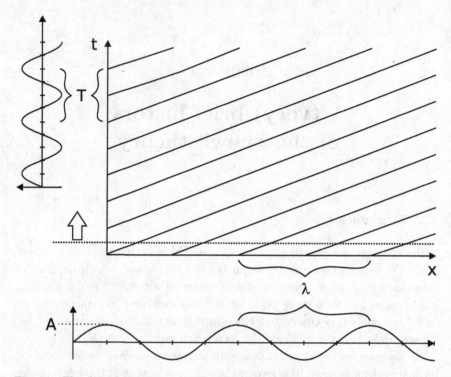

Fig. 4.1   The slanted lines represent the alternating crests and troughs of $\psi(x,t)$. The passing of time is indicated by the upward-moving dotted line, which stands for the temporal present. As time passes, the crests and troughs move toward the right. By focusing on a fixed time one can see that a cycle (crest to crest, say) completes after a distance λ. By focusing on a fixed place, one can see that a cycle completes after a time T.

the origin of this relation in Sec. 9.3.) In the presence of a potential energy $E_P(t,x)$, the electron's total energy is

$$E = \frac{p^2}{2m} + E_P.$$

Multiplying this equation with ψ,

$$E\psi = \frac{p^2}{2m}\psi + E_P\,\psi,$$

and using Eqs. (4.3) to substitute $i\hbar\,(\partial\psi/\partial t)$ for $E\psi$ and $-\hbar^2(\partial^2\psi/\partial x^2)$ for $p^2\,\psi$, we arrive at the 1-dimensional *Schrödinger equation*,

$$i\hbar\frac{\partial\psi}{\partial t} = -\frac{\hbar^2}{2m}\frac{\partial^2\psi}{\partial x^2} + E_P\,\psi. \tag{4.4}$$

The generalization to three dimensions is straightforward:

$$i\hbar\frac{\partial\psi}{\partial t} = -\frac{\hbar^2}{2m}\left(\frac{\partial^2\psi}{\partial x^2} + \frac{\partial^2\psi}{\partial y^2} + \frac{\partial^2\psi}{\partial z^2}\right) + E_P\,\psi\,. \tag{4.5}$$

Let us however stick to the 1-dimensional case.

Problem 4.1. *If $f(x,t)$ and $g(x,t)$ are functions that satisfy Eq. (4.4), then Eq. (4.4) is also satisfied by the function $h = af + bg$, where a and b are complex constants.*

Our derivation of Eq. (4.4) ensures that its free version ($E_P = 0$) is satisfied by any function of the form (4.2) for which $E = p^2/2m$. The general solution of the free Schrödinger equation is therefore[1]

$$\psi(x,t) = \frac{1}{\sqrt{2\pi}}\int_{-\infty}^{+\infty}\tilde{\psi}(k)\,e^{i[kx-\omega(k)t]}\,dk\,, \tag{4.6}$$

where $\tilde{\psi}(k)$ is the (complex) wave amplitude associated with

$$\psi_k(x,t) \stackrel{\text{Def}}{=} \frac{1}{\sqrt{2\pi}}\,e^{i[kx-\omega(k)t]}\,, \tag{4.7}$$

and $\omega(k) = \hbar k^2/2m$. If we now define

$$\overline{\psi}(k,t) \stackrel{\text{Def}}{=} \tilde{\psi}(k)\,e^{-i\omega(k)t}\,, \tag{4.8}$$

we have that

$$\psi(x,t) = \frac{1}{\sqrt{2\pi}}\int_{-\infty}^{+\infty}\overline{\psi}(k,t)\,e^{ikx}\,dk\,, \tag{4.9}$$

and this tells us that $\psi(x,t)$ is the *Fourier transform* of $\overline{\psi}(k,t)$. As a consequence, we also have that

$$\overline{\psi}(k,t) = \frac{1}{\sqrt{2\pi}}\int_{-\infty}^{+\infty}\psi(x,t)\,e^{-ikx}\,dx\,. \tag{4.10}$$

4.2   Born

Now here's the one million dollar question: what does the Schrödinger equation have to do with the real world? What do its solutions tell us about the physical systems with which they are associated?

In the same year that Schrödinger published the equation that now bears his name, the non-relativistic theory was completed by Max Born's insight

---

[1]You will learn the reason for including the factor $1/\sqrt{2\pi}$ by doing Problem 11.2.

that the "wave function" $\psi(x, y, z, t)$ is *a tool for calculating probabilities of measurement outcomes*. Specifically, if ψ is associated with a particle, its absolute square $|\psi(x, y, z, t)|^2$ is a (time-dependent) *probability density*, in the sense that the probability of detecting the particle in a region R, by a measurement made at the time t, is given by

$$p(R, t) = \int_R |\psi(x, y, z, t)|^2 \, d^3r \,. \tag{4.11}$$

Since the probability of finding the particle *somewhere* (no matter where) has to be equal to unity, a meaningful solution of the Schrödinger equation must be *square-integrable*: the integral of its absolute value over any interval or region must stay finite in the limit that the length of the interval or the volume of the region tends to infinity.

For a continuous variable x with a normalized probability density $\rho(x)$,[2] the expected value (1.11) and the standard deviation (1.12) take the respective forms

$$\langle x \rangle = \int_{-\infty}^{+\infty} \rho(x) \, x \, dx \,, \tag{4.12}$$

$$\Delta x = \sqrt{\int_{-\infty}^{+\infty} \rho(x)(x - \langle x \rangle)^2 \, dx} \,. \tag{4.13}$$

Thus if $\psi(x, t)$ is a solution of the 1-dimensional Schrödinger equation, then

$$\langle x(t) \rangle = \int_{-\infty}^{+\infty} |\psi(x, t)|^2 \, x \, dx \tag{4.14}$$

is the expected value associated with a position measurement made at the time t, and

$$\Delta x(t) = \sqrt{\int_{-\infty}^{+\infty} |\psi(x, t)|^2 \big(x - \langle x(t) \rangle\big)^2 \, dx} \tag{4.15}$$

is the corresponding standard deviation. By the same token,[3]

$$\langle p(t) \rangle = \hbar \, \langle k(t) \rangle = \hbar \int_{-\infty}^{+\infty} |\overline{\psi}(k, t)|^2 \, k \, dk \tag{4.16}$$

is the expected value associated with a momentum measurement made at the time t, and

$$\Delta p(t) = \hbar \, \Delta k(t) = \hbar \sqrt{\int_{-\infty}^{+\infty} |\overline{\psi}(k, t)|^2 \big(k - \langle k(t) \rangle\big)^2 \, dk} \tag{4.17}$$

is the corresponding standard deviation.

---

[2] "Normalized" here means that $\int_{-\infty}^{+\infty} \rho \, dx = 1$.

[3] The following statements are corroborated in Secs. 11.3 and 11.6.

4.3   Heisenberg and "uncertainty"

Because $\psi(x, t)$ and $\overline{\psi}(k, t)$ are Fourier transforms of each other, the standard deviations Δx and Δk are constrained by the inequality $\Delta x \, \Delta k \geq 1/2$. It follows that

$$\Delta x \, \Delta p \geq \frac{\hbar}{2} \, . \tag{4.18}$$

This "uncertainty relation," as it is generally called in the English speaking world, was first derived by Werner Heisenberg, also in 1926. It is essential to our understanding of why the laws of quantum mechanics have the form that they do.

Bohr, as you will remember, postulated the quantization of angular momentum in an effort to explain the stability of atoms. An atom "occupies" hugely more space than its nucleus (which is *tiny* by comparison) or any one of its electrons (which do not appear to "occupy" any space at all). How then does an atom come to "occupy" as much space as it does without collapsing? The answer is: because of the "uncertainties" in both the positions and the momenta of its electrons relative to its nucleus. It is these "uncertainties" that "fluff out" matter.

Except that "uncertainty" cannot then be the right word. What "fluffs out" matter cannot be our very own, subjective *uncertainty* about the values of the relative positions and momenta of its constituents. It has to be an objective *fuzziness* of these values.

Consider again the lowly hydrogen atom. It seems clear enough that the fuzziness of the electron's position relative to the proton can be at least partly responsible for the amount of space that the atom "occupies." (For a hydrogen atom in its ground state, this is a space roughly one tenth of a nanometer across.) But *being* fuzzy is not enough. This position must also *stay* fuzzy, and that is where the fuzziness of the corresponding momentum comes in.

The standard deviation Δr associated with the radial component r of the electron's position relative to the nucleus is a measure of the fuzziness of r. If the electrostatic attraction between the electron and the proton were the only force at work, it would cause a decrease in Δr, and the atom would collapse as a result. The stability of the atom requires that the electrostatic attraction be counterbalanced by an effective repulsion. In view of the fact that we already have a fuzzy relative position, the absolutely simplest way of obtaining such a repulsion—and a darn elegant way at that—is to let the corresponding relative momentum be fuzzy, too. As Fig. 4.2 illustrates,

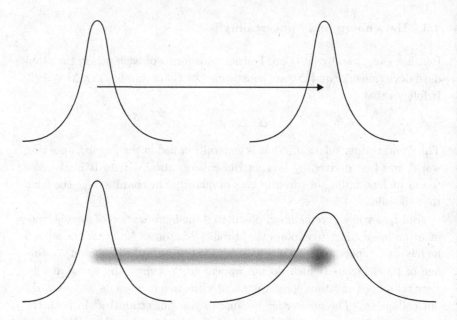

Fig. 4.2 A fuzzy position, represented by a probability density function, at two different times. Above: if an object with this position has an exact momentum, it moves by an exact distance; the fuzziness of its position therefore does not increase. Below: if the same object has a fuzzy momentum, it moves by a fuzzy distance; as a result, its position grows fuzzier.

a fuzzy momentum causes a fuzzy position to grow more fuzzy. If the electrostatic attraction were absent, the fuzziness of the momentum would causes an increase in Δr. In its presence, equilibrium is possible.

But if a *stable* equilibrium is to be maintained, more is needed. If the mean distance between the two particles decreases, their electrostatic attraction increases. A stable equilibrium is possible only if the effective repulsion increases at the same time. We therefore expect a decrease in Δr to be accompanied by an increase in Δp, the fuzziness of the radial component of the momentum, and we expect an increase in Δr to be accompanied by a decrease in Δp. We therefore expect the product $\Delta r\, \Delta p$ to have a positive lower limit. The atom's stability implies a relation of the form (4.18).

Let us work out the speeds at which fuzzy positions get fuzzier when left to themselves. To this end we consider an object associated with the Gaussian wave function

$$\psi(x,0) = \frac{1}{\sqrt{\sigma\sqrt{\pi}}} e^{-x^2/2\sigma^2}. \tag{4.19}$$

The factor in front of the exponential ensures that the probability density

$$|\psi(x,0)|^2 = \frac{1}{\sigma\sqrt{\pi}} e^{-x^2/\sigma^2} \tag{4.20}$$

is normalized. The standard deviation of $|\psi(x,0)|^2$ works out at

$$\Delta x(0) = \sigma/\sqrt{2}. \tag{4.21}$$

To find out how this grows with time, we calculate the Fourier transform of $\psi(x,0)$ using Eq. (4.10). The result is

$$\overline{\psi}(k,0) = \sqrt{\frac{\sigma}{\sqrt{\pi}}} e^{-\sigma^2 k^2/2}.$$

This defines the normalized probability density

$$|\overline{\psi}(k,0)|^2 = \frac{\sigma}{\sqrt{\pi}} e^{-\sigma^2 k^2}$$

with standard deviation $\Delta k(0) = 1/\sigma\sqrt{2}$. Observe that the initial uncertainties associated with the object's position and momentum satisfy the lower bound of the inequality (4.18): $\Delta x(0)\,\Delta p(0) = \hbar/2$.

The final step is to calculate the Fourier transform of $\overline{\psi}(k,t)$ with the help of Eq. (4.9), having gathered from Eq. (4.8) that $\overline{\psi}(k,t) = \overline{\psi}(k,0)\,e^{-i\omega(k)t}$. The result defines a probability density with standard deviation

$$\Delta x(t) = \sqrt{\frac{\sigma^2}{2} + \frac{\hbar^2 t^2}{2m^2\sigma^2}}. \tag{4.22}$$

Equation (4.21) allows us to cast this into the final form

$$\Delta x(t) = \sqrt{[\Delta x(0)]^2 + \frac{\hbar^2 t^2}{4m^2[\Delta x(0)]^2}}. \tag{4.23}$$

Two factors thus impact the speed with which the fuzziness Δx of our object's position spreads: the initial fuzziness $\Delta x(0)$ and the object's mass m. The smaller they are, the faster Δx grows. Suppose, for example, that $\Delta x(0)$ is a tenth of a nanometer (a typical size for an atom). If m is the mass of an electron, Δx will grow at a whopping $600\,\mathrm{km/s}$. If m is the mass of a C_{60} molecule, Δx will grow at a moderate $44\,\mathrm{cm/s}$. And if m is the mass of a peanut, Δx will grow at a rate of about $2 \times 10^{-24}\,\mathrm{m/s}$; it will take the present age of the universe to grow to about 750 nanometers.

This illustrates one of the reasons why the fuzziness of the positions of everyday objects is not readily observed.

4.4   Why energy is quantized

We have seen how the "old" theory accounts for the quantization of energy. Let us now see how the "new" theory does it. We begin by observing that if E_P does not depend on time, then the Schrödinger equation (4.5) has solutions that are products of a time-independent function $\psi(x,y,z)$ and a time-dependent phase factor $e^{-(i/\hbar)\,E\,t}$:

$$\Psi(x,y,z,t) = \psi(x,y,z)\,e^{-(i/\hbar)\,E\,t}. \tag{4.24}$$

Because the probability density $|\Psi|^2$ is independent of time, these solutions are called *stationary*.

Problem 4.2. *Both $\langle k \rangle$ and Δk are constant.*

Let us now find the solutions of Eq. (4.26) that can be associated with a particle trapped inside a potential well like that in Fig. 4.3. Between x_1 and x_2, the particle's total energy E exceeds E_P, so that $\psi(x)$ exhibits wavelike behavior. To the left of x_1 and to the right of x_2, E is smaller than E_P, so that $\psi(x)$ curves away from the x axis. (A classical particle would oscillate to and fro between these two points.) The only way to obtain a square-integrable solution is to make sure that the graph of $\psi(x)$ approaches the x axis asymptotically in both the limits $x \to +\infty$ and $x \to -\infty$. But for this the value of E must be exactly right.

Figure 4.4 sketches the first six solutions. The first solution lacks nodes. Decreasing E below the ground state energy E_0 will not yield another solution. Increasing E just a little will not yield a solution either. E must increase by a specific amount, from E_0 to E_1, if another solution is to be obtained. The possible ("allowed") energies thus form a sequence E_n, with $n \geq 0$ counting the number of nodes.

Let us remind ourselves, in concluding this chapter, what exactly these stationary states represent. Every quantum-mechanical probability algorithm serves a single purpose: to calculate the probabilities of possible measurement outcomes on the basis of actual outcomes. If a particle is associated with one of these states, the situation envisaged is one in which an energy measurement has been made and an outcome E_m has been obtained. The corresponding wave function $\psi_m(x)$ allows us to calculate

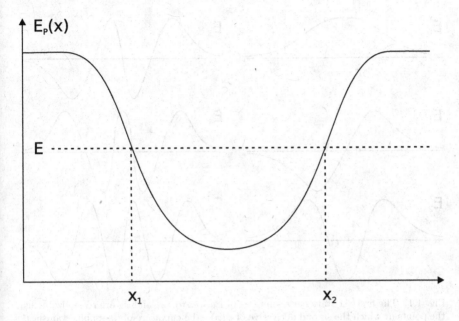

Fig. 4.3 A potential well.

the probability of finding the particle's position in any given interval of the x axis, provided that the appropriate measurement is made, while the Fourier transform of $\psi_m(x)$ allows us to calculate the probability of finding the particle's momentum in any given interval of the k axis—again provided that the appropriate measurement is made.

Problem 4.3. *Plug the function (4.24) into Eq. (4.5) to find that $\psi(x, y, z)$ satisfies the* time-independent Schrödinger equation

$$E\,\psi = -\frac{\hbar^2}{2m}\left(\frac{\partial^2\psi}{\partial x^2} + \frac{\partial^2\psi}{\partial y^2} + \frac{\partial^2\psi}{\partial z^2}\right) + E_P\,\psi. \tag{4.25}$$

We now cast the 1-dimensional version of this equation into the form

$$\frac{d^2\psi(x)}{dx^2} = A(x)\,\psi(x)\,, \quad \text{where} \quad A(x) = \frac{2m}{\hbar^2}\Big[E_P(x) - E\Big]. \tag{4.26}$$

Because Eq. (4.26) does not contain any complex numbers (apart from, possibly, ψ itself), it has real-valued solutions. So let us assume that $\psi(x)$ is real. The first thing we notice is that if $E_P > E$ then $\psi(x)$ has the same sign as its second derivative. This means that the slope of $\psi(x)$ increases above and decrease below the x axis, so that $\psi(x)$ crosses this axis at most

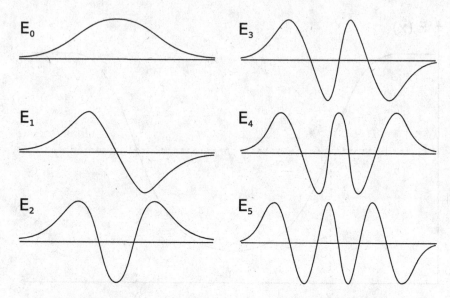

Fig. 4.4   The first six stationary solutions of Eq. (4.26). x_1 and x_2 are recognizable as the points at which the second derivative of $\psi(x)$—the curvature of its graph—vanishes. At x_1, the graph changes from bending away from the axis to bending toward it, and at x_2 it changes back from bending toward the axis to bending away from it.

once. On the other hand, if $E_P > E$ then $\psi(x)$ has the opposite sign of its second derivative. This means that the graph of $\psi(x)$ bends toward the x axis and keeps crossing it—just like a wave.

Chapter 5

The Feynman route to Schrödinger (stage 1)

5.1 The rules of the game

Suppose that you want to calculate the probability of a particular outcome of a measurement M_2 given the outcome of an earlier measurement M_1. Here is what you have to do:

- Choose a sequence of measurements that may be made in the meantime.
- Assign to each possible sequence of intermediate outcomes (called "alternative") a complex number (called "amplitude").
- Apply the appropriate rule:

  (A) If the intermediate measurements are made (or if it is possible to find out what their outcomes would have been if they had been made), first square the absolute values of the amplitudes of the alternatives and then add the results.

  (B) If the intermediate measurements are not made (and if it is impossible to find out what their outcomes would have been if they had been made), first add the amplitudes of the alternatives and then square the absolute value of the result.

These rules will be derived in Chap. 8. Here we will use them as our starting point.

5.2 Two slits

According to Feynman *et al.* (1965), the following experiment "has in it the heart of quantum mechanics." The setup consists of an electron gun G, a plate with two slits L and R equidistant from G, and a backdrop or

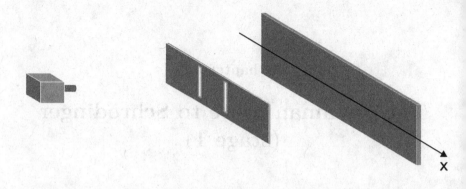

Fig. 5.1   Setup for the two-slit experiment with electrons.

screen at which electrons are detected (Fig. 5.1). The initial measurement indicates that an electron has been launched at (the position of) G. (If we assume, as we will, that G is the only source of free electrons, then the detection of an electron behind the slit plate also indicates the electron's launch at G.) The final measurement indicates the position—along an axis across the backdrop—at which the electron is detected. A single intermediate measurement, if made, indicates the slit through which the electron went. Thus there are two alternatives:

- The electron went through the left slit (L).
- The electron went through the right slit (R).

The corresponding amplitudes will be denoted by a_L and a_R. The event whose probability we wish to calculate is the detection of the electron by a detector (situated at) D. Here is what we need to know in order to be able to perform this calculation:

- a_L is the product of two complex numbers, called *propagators*, for which we shall use the symbols $\langle D|L \rangle$ and $\langle L|G \rangle$. Thus $a_L = \langle D|L \rangle \langle L|G \rangle$. By the same token, $a_R = \langle D|R \rangle \langle R|G \rangle$.
- The absolute value of $\langle B|A \rangle$ is inverse proportional to the distance $\overline{BA}$ between A and B.
- The phase of $\langle B|A \rangle$ is proportional to $\overline{BA}$.

5.2.1   Why product?

In Sec. 4.3 we arrived at the conclusion—without invoking the "uncertainty" relation (4.18)—that the stability of atoms requires the product

$\Delta x \, \Delta p$ to have a positive lower limit. Hence $\Delta p \to \infty$ as $\Delta x \to 0$. What does this tell us about the probability $p(C \leftarrow B | A)$ with which an electron launched at A is detected first at B and then at C—assuming for simplicity's sake that A, B, and C are exact locations?

If the electron is detected at B, its momentum at the time of detection at B is as fuzzy as it gets. Hence the probability $p(C | B)$, with which an electron that has been detected at B is detected at C, does not depend on whether or not it was initially launched at A. The probability $p(C | B)$ of its detection at C (given that it has been detected at B) and the probability $p(B | A)$ of its detection at B (given that it has been launched at A) are therefore independent. The product rule of Sec. 1.4 thus applies: $p(C \leftarrow B | A) = p(C | B) \, p(B | A)$. And since the probability associated with a single alternative is the absolute square of its amplitude, the product rule also applies to the amplitude $a(C \leftarrow B | A)$; this is the product of $a(C | B) = \langle C | B \rangle$ and $a(B | A) = \langle B | A \rangle$.

5.2.2 Why inverse proportional to $\overline{BA}$?

Imagine a sphere of radius R centered at A. Let p_R be the probability with which a particle launched at A is found by a detector monitoring a unit area of the surface S of this sphere. Because the particle proceeds from A in no particular direction, p_R is constant across S. If S is covered with detectors that jointly monitor the entire sphere, the probability that one of them—no matter which—detects the particle, equals 1. Since (the area of) S is proportional to R^2, this means that p_R is inverse proportional to R^2, and that the corresponding amplitude is inverse proportional to R. Hence $|\langle B | A \rangle|$ is inverse proportional to $\overline{BA}$.

5.2.3 Why proportional to $\overline{BA}$?

Because a particle launched at A proceeds in no particular direction, the phase of $\langle B | A \rangle$ cannot depend on the direction in which B lies relative to A. It can only depend on the distance between A and B. And if we use a sensible metric, equal phases will correspond to equal distances.

A *metric* is a method of assigning distances to pairs of points or, more generally, a method of assigning lengths to curves. It is one of those mathematical machines: insert two points or insert a curve, and out pops the distance between the points or the length of the curve. What is important here is that space does not come with an inbuilt metric. It is *we* who chose

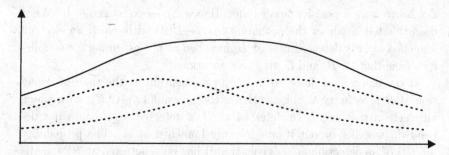

Fig. 5.2   The probability of detection according to Rule A (the solid curve) is the sum of two probability distributions (the dotted curves), one for the electrons that went through L and one for the electrons that went through R.

the metric that we shall use, and our choice ought to be such as to make the formulation of the laws of physics as simple as possible. At the basic level of propagators associated with particles moving freely, the obvious choice is to let equal distances correspond to equal phases. The phase of $\langle B|A \rangle$ will therefore be proportional to $\overline{BA}$.

5.3 * Interference

We are now in a position to calculate the probability $p(D|G)$ with which an electron launched at G is found by a detector situated at D. According to Rule A (Fig. 5.2),

$$p_A(D|G) = \left| \langle D|L \rangle \langle L|G \rangle \right|^2 + \left| \langle D|R \rangle \langle R|G \rangle \right|^2. \tag{5.1}$$

The slits are equidistant from G, so $\langle L|G \rangle = \langle R|G \rangle$. Hence in order to be able to plot $p_A(D|G)$ as a function of the position x of D, we only need to calculate

$$|\langle D|L \rangle|^2 + |\langle D|R \rangle|^2 = \frac{1}{\overline{DL}^2} + \frac{1}{\overline{DR}^2}. \tag{5.2}$$

The missing overall factor can be recovered by requiring that the area under the plot should be 1.

Calculated according to Rule B (Fig. 5.3), the probability $p_B(D|G)$ with which an electron launched at G is found by a detector at D is proportional to $|\langle D|L \rangle + \langle D|R \rangle|^2$. This equals

$$\left(\frac{e^{ik\overline{DL}}}{\overline{DL}} + \frac{e^{ik\overline{DR}}}{\overline{DR}} \right) \left(\frac{e^{-ik\overline{DL}}}{\overline{DL}} + \frac{e^{-ik\overline{DR}}}{\overline{DR}} \right) = \frac{1}{\overline{DL}^2} + \frac{1}{\overline{DR}^2} + \frac{2\cos(k\Delta)}{\overline{DL}\,\overline{DR}},$$

$$\tag{5.3}$$

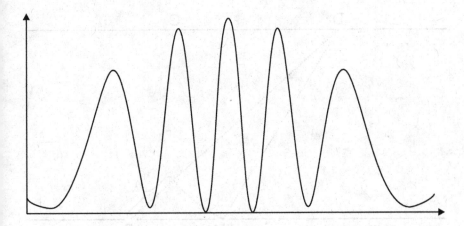

Fig. 5.3   The probability of detection according to Rule B.

where Δ is the difference between $\overline{DR}$ and $\overline{DL}$, and k is the proportionality factor that relates the phase $\alpha(\overline{BA}) = k\,\overline{BA}$ to the distance $\overline{BA}$.

Problem 5.1. ($*$) *Verify these calculations.*

The last term of Eq. (5.3), which is responsible for the difference between the plots in Figs. 5.2 and 5.3, is frequently referred to as an "interference term." It accounts for two remarkable facts. The first is that near the (local) minima of Fig. 5.3, the probability of detection is *less* if both slits are open than if one slit is shut. The second is that the (absolute) maximum of Fig. 5.3 is twice as high as the maximum of Fig. 5.2. The first fact is usually attributed to "destructive interference," the second to "constructive interference." This terminology is prone to cause a great deal of confusion. Quantum-mechanical interference is *not* a physical mechanism or process. Whenever we say that "constructive interference occurs," we simply mean that a probability calculated according to Rule B is greater than the same probability calculated according to Rule A. And whenever we say that "destructive interference occurs," we simply mean that a probability calculated according to Rule B is less than the same probability calculated according to Rule A.

5.3.1   Limits to the visibility of interference fringes

The only quantity that is characteristic of the freely propagating particles used in this experiment is the proportionality factor k. As was mentioned

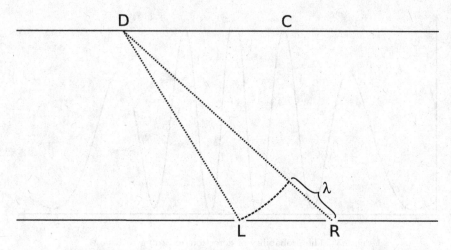

Fig. 5.4   If the central maximum of Fig. 5.3 is located at C and the difference between $\overline{RD}$ and $\overline{LD}$ equals λ, then the first maximum to the left of C is located at D.

in Sec. 5.2.3, its constancy ensures that equal distances correspond to equal phases. Because phases are cyclic with a period of 2π, the particles in this experiment define a unit of length λ via $k = 2\pi/\lambda$. While distances *per se*, unlike meter sticks, are unobservable, this unit of length can be deduced from the interference pattern produced by the particles. Figure 5.4 shows how.

But the wavenumber $k = 2\pi/\lambda$ associated with a particle is related to the particle's momentum via $p = \hbar k$, and p is proportional to the particle's mass. If the two-slit experiment is performed with particles of greater mass (other things being equal), λ will be smaller, and the interference maxima will be more closely spaced as a result (Fig. 5.5). If we keep increasing the mass of the particles employed, there comes a point beyond which the maxima can no longer be resolved, for the intervals monitored by real detectors cannot be made arbitrarily small. Several maxima and minima will come to lie in the same interval I, so that the relative frequency with which particles are detected in I will indicate the mean value of $p_B(D|G)$ across I. A plot obtained with insufficiently small detectors will therefore resemble the solid line in Fig. 5.2. The peculiar behavior predicted by Rule B will no longer be in evidence. This is another reason why everyday objects appear to behave in accordance with the laws of classical physics.

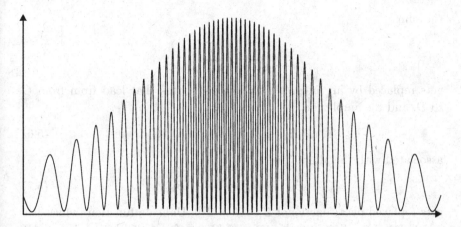

Fig. 5.5   An interference pattern obtained with particles of greater mass.

The limit imposed on the observability of interference patterns by the mass of the objects employed can be raised by using diffraction gratings—plates containing large numbers of closely spaced slits—instead of a plate with only two slits [see Arndt *et al.* (1999); Brezger *et al.* (2002); Nairz *et al.* (2001)]. But the necessity of using closely spaced and hence *narrow* slits then limits the size of the objects with which interference can be observed.

5.4   The propagator as a path integral

Let us replace the plate with two slits by a plate with n holes B_k. Instead of being the sum of two terms, the propagator $\langle D|G\rangle$ is now a sum of n terms,

$$\langle D|G\rangle = \sum_{k=1}^{n} \langle D|B_k\rangle \langle B_k|G\rangle. \tag{5.4}$$

Next, let us replace the plate with n holes by m such plates a distance Δy apart. The resulting propagator is

$$\langle D|G\rangle = \sum_{k_1=1}^{n} \cdots \sum_{k_m=1}^{n} \langle D|B_{k_m}\rangle \cdots \langle B_{k_2}|B_{k_1}\rangle \langle B_{k_1}|G\rangle. \tag{5.5}$$

Let us drill more holes in each plate, and let us add more plates. And still more holes, and still more plates. What happens if we drill so many holes in each plate that the plates are no longer there? What happens is that the sum

$$\sum_{k_1=1}^{n} \cdots \sum_{k_m=1}^{n}$$

gets replaced by an integral $\int \mathcal{DC}$ over all paths that lead from from G to D, and the amplitude

$$\langle D|B_{k_m}\rangle \cdots \langle B_{k_2}|B_{k_1}\rangle \langle B_{k_1}|G\rangle \tag{5.6}$$

associated with the alternative

$$G \to B_{k_1} \to B_{k_2} \to \cdots \to B_{k_m} \to D \tag{5.7}$$

gets replaced by a functional $Z[\mathcal{C}, G \to D]$. Unlike a function, which has input slots for a finite number of variables, a *functional* has an input slot for a *function*—in this case the function that describes the path $\mathcal{C}$ from G to D. The resulting propagator is

$$\langle D|G\rangle = \int \mathcal{DC}\, Z[\mathcal{C}, G \to D]. \tag{5.8}$$

This is not an ordinary (Riemann) integral $\int_a^b dx\, f(x)$, to which each infinitesimal interval dx makes a contribution proportional to the value of $f(x)$ inside it, but a *path integral*, to which each infinitesimal bundle of paths $\mathcal{DC}$ makes a contribution proportional to the value of Z inside it. As it stands, it is no more than an idea of an idea, albeit a fruitful one, as we shall see.

Let us note, to begin with, that if a path $\mathcal{C}$ consists of two segments $\mathcal{C}_1$ and $\mathcal{C}_2$, the multiplicativity of successive propagators, Eq. (5.4), translates to

$$Z[\mathcal{C}] = Z[\mathcal{C}_2]\, Z[\mathcal{C}_1]. \tag{5.9}$$

5.5 The time-dependent propagator

The alternatives considered so far in this chapter were independent of time. It did not matter when the electron left G, when it passed the slit plate, or when it was detected at D. The measurements considered were measurements that indicated (or would have indicated if performed) *where* particles were detected but not *when* they were detected. The paths considered in the previous section were paths in space, not paths in *spacetime* (Fig. 5.6).

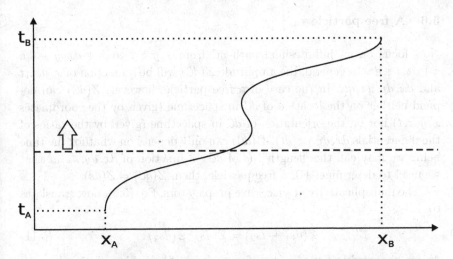

Fig. 5.6   A typical spacetime diagram displaying one space axis and the time axis. Also shown is the spacetime path $\mathcal{C}$ of a (classical) object that travels along the x axis with varying speed. It leaves x_A at the time t_A and arrives at x_B at the time t_B. Its speed dx/dt is the inverse of the slope dt/dx of $\mathcal{C}$. The upward motion of the dashed line suggests time's passage toward the future.

Problem 5.2. *Describe the motion of the object whose spacetime path is shown in Fig. 5.6. Where does it accelerate, where does it decelerate, where does it reverse its direction of motion?*

Let us now interpose, between the particle's launch at G and its detection at D, a succession of m position measurements made at regular intervals Δt. Each measurement uses an array of detectors monitoring n mutually disjoint regions $R_1, R_2, \ldots R_n$. In other words, we replace the plates, each extending in two spatial dimensions and stacked into the third, by hyperplanes extending in all three spatial dimensions and stacked into the temporal dimension, and we replace the spacing Δy of the plates by the time Δt between measurements. What happens in the limit in which (i) the total region monitored becomes the whole of space, (ii) the volumes of the individual regions tend to zero, and (iii) Δt also tends to zero?

What happens is that the integral (5.8) over paths in space leading from G to D becomes an integral over paths in spacetime that start at the point $\mathbf{r}_A$ (with coordinates x_A, y_A, z_A) at the time t_A and end at the point $\mathbf{r}_B$ (with coordinates x_B, y_B, z_B) at the time t_B:

$$\langle \mathbf{r}_B, t_B | \mathbf{r}_A, t_A \rangle = \int \mathcal{DC}\, Z[\mathcal{C}|\mathbf{r}_A, t_A \to \mathbf{r}_B, t_B]. \tag{5.10}$$

5.6   A free particle

Now focus on an infinitesimal path $d\mathcal{C}$ from x, y, z, t to $x + dx, y + dy$, $z + dz, t + dt$. In general, the amplitude $Z(d\mathcal{C})$ will be a function of x, y, z, t and dx, dy, dz, dt. In the case of a free particle, however, $Z(d\mathcal{C})$ can depend neither on the location of $d\mathcal{C}$ in spacetime (given by the coordinates x, y, z, t) nor on the orientation of $d\mathcal{C}$ in spacetime (given by the ratios of the differentials dx, dy, dz, dt). $Z(d\mathcal{C})$ can only depend on what for the time being we may call the "length" ds of $d\mathcal{C}$—a function of dx, dy, dz, dt that we need to determine. For a free particle, then, $Z(d\mathcal{C}) = Z(ds)$.

The multiplicativity of successive propagators, Eq. (5.9), now translates to

$$Z(ds_1 + ds_2) = Z(ds_1) Z(ds_2). \tag{5.11}$$

As we observed in Sec. 3.5, any function satisfying this equation has the form $f(x) = a^x$. It follows that $Z(ds)$ is of the form a^{ds} or, if we introduce a complex number z such that $a = e^z$,

$$Z(ds) = e^{z\,ds}. \tag{5.12}$$

Equation (5.11) now takes the form

$$e^{z(ds_1 + ds_2)} = e^{z\,ds_1} e^{z\,ds_2}. \tag{5.13}$$

If instead of combining two infinitesimal path segments we combine all infinitesimal segments that make up a spacetime path $\mathcal{C}$, we obtain the amplitude associated with $\mathcal{C}$:

$$Z[\mathcal{C}] \overset{(1)}{=} \prod_{\mathcal{C}} Z(d\mathcal{C}) \overset{(2)}{=} \prod_{\mathcal{C}} Z(ds) \overset{(3)}{=} \prod_{\mathcal{C}} e^{z\,ds} \overset{(4)}{=} e^{z \int_{\mathcal{C}} ds} \overset{(5)}{=} e^{zs[\mathcal{C}]}. \tag{5.14}$$

The first equality generalizes Eq. (5.9). The product $\prod_{\mathcal{C}}$ multiplies the infinitely many amplitudes associated with the infinitesimal segments that make up $\mathcal{C}$. The second equality holds for a free particle. The third follows from Eq. (5.12). The fourth generalizes Eq. (5.13). The integral $\int_{\mathcal{C}} ds$, which adds up the "lengths" of the infinitesimal segments that make up $\mathcal{C}$, equals the "length" $s[\mathcal{C}]$ of $\mathcal{C}$; this accounts for the fifth equality.

5.7   A free and stable particle

We proceed to calculate the probability with which a free particle detected at $\mathbf{r}_A$ at the time t_A still exists at the time t_B. For a *stable* free particle this must be equal to unity. We obtain this probability by integrating the absolute square of the propagator (5.10) (as a function of $\mathbf{r}_B$) over the whole of space, with $Z[\mathcal{C}]$ replaced by $e^{z\,s[\mathcal{C}|\mathbf{r}_A,t_A\to\mathbf{r}_B,t_B]}$:

$$\int |\langle \mathbf{r}_B,t_B|\mathbf{r}_A,t_A\rangle|^2\, d^3r_B = \int \left| \int \mathcal{DC}\, e^{z\,s[\mathcal{C}|\mathbf{r}_A,t_A\to\mathbf{r}_B,t_B]} \right|^2 d^3r_B. \qquad (5.15)$$

If you contemplate this with a calm heart and an open mind, you will notice that if the complex number $z = a + ib$ had a real part $a \neq 0$, the path integral would contain a factor $e^{a\,s[\mathcal{C}|\mathbf{r}_A,t_A\to\mathbf{r}_B,t_B]}$, and the integral over space, considered as a function of t_B, would blow up exponentially (if $a > 0$) or fall off exponentially (if $a < 0$). It could not remain equal to unity. For a free and stable particle, therefore, z is an imaginary number ib, and the amplitude associated with a path $\mathcal{C}$ is

$$Z[\mathcal{C}|\mathbf{r}_A,t_A \to \mathbf{r}_B,t_B] = e^{ibs[\mathcal{C}|\mathbf{r}_A,t_A\to\mathbf{r}_B,t_B]}. \qquad (5.16)$$

The behavior of a free and stable particle thus is controlled by a single real number b. To discover its significance, we need to know more about the "length" functional $s[\mathcal{C}]$, and to this end we need to acquaint ourselves with the special theory of relativity.

Chapter 6

Special relativity in a nutshell

6.1   The principle of relativity

Let us imagine a universe containing a single object. Would we be able to attribute to this object a position—to say *where* it is? Of course we wouldn't. At a minimum we need *two* objects for this. If we have two objects, as well as a method of measuring their distance, we can attribute to them a distance, so we can say how far one is from the other. We can imagine a straight line from one to the other, but if this is all there is, we cannot attribute to this line an *orientation*. The bottom line: there is no such thing as an *absolute* position or orientation. Positions and orientations are *relative*. If we want to make physical sense, we can only speak of the positions and orientations of physical objects *relative to* other physical objects.

By the same token, if we were to imagine a world in which a single event takes place, we would not be able to attribute to this event a time—to say *when* it occurs. At a minimum we need *two* events for this. If we have two events, as well as a method of measuring the time interval between them, we can say how much time has passed between them. So there is no such thing as an *absolute* time either. If we want to make physical sense, we can only speak of the times of physical events *relative to* other physical events.[1]

And if we were to again imagine a world containing a single object, we wouldn't be able to attribute to this object a velocity—to say how fast it

---

[1]While one may think of the Big Bang as the beginning of time, this singular event transcends the *local* physics we are concerned with at present. "Local" is used here in the same sense in which we say that a curve is *locally* straight or a warped surface is *locally* flat—the curvature of a sufficiently small segment of a line or patch of a surface can be ignored. While spacetime itself may be and, in a sense, *is* warped, the special theory of relativity deals with regions of spacetime that are so small or so weakly warped that their curvature need not be taken into account.

was moving, in which direction it was moving, or even if it was moving at all. So there is no such thing as *absolute* rest either. If we want to make physical sense, we can only speak of the velocities with which physical objects move *relative to* each other.

All of this is encapsulated in the *principle of relativity*, according to which *all inertial coordinate systems are created equal*: the laws of physics do not "favor" any particular inertial frame or class of such frames.

A coordinate system or frame is *inertial* if (and only if) the components x, y, z of the position of any freely moving classical object change by equal amounts $\Delta x, \Delta y, \Delta z$ in equal time intervals Δt. In other words, the ratios formed of $\Delta x, \Delta y, \Delta z$ and Δt are constants. Hence if $\mathcal{F}$ is an inertial frame, then so is any coordinate frame that, relative to $\mathcal{F}$,

- is shifted ("translated") in space by a given distance in a given direction,
- is shifted ("translated") in time by a given amount of time,
- is rotated by a given angle about a given axis, and/or
- moves with a constant velocity.

6.2 Lorentz transformations: General form

The task now before us is to express the coordinates t and $\mathbf{r} = (x, y, z)$ of an inertial frame $\mathcal{F}_1$ in terms of the coordinates t' and $\mathbf{r}' = (x', y', z')$ of another inertial frame $\mathcal{F}_2$. We will make the following assumptions:

- The coordinate origins of the two frames coincide:
  $t' = 0$ and $\mathbf{r}' = 0$ mark the same spacetime location as $t = 0$ and $\mathbf{r} = 0$.
- The spatial axes of the two frames are parallel.
- $\mathcal{F}_2$ moves with a constant velocity $\mathbf{w}$ relative to $\mathcal{F}_1$.

Whatever moves with a constant velocity in one inertial frame will do so in any other inertial frame. This means that the transformation from $t, \mathbf{r}$ to $t', \mathbf{r}'$ maps straight lines onto straight lines (in spacetime). Such a transformation is linear: the dashed coordinates t', x', y', z' are *linear combinations* of the undashed ones:

$$t' = a_{00}t + a_{01}x + a_{02}y + a_{03}z \,,$$
$$x' = a_{10}t + a_{11}x + a_{12}y + a_{13}z \,,$$
$$y' = a_{20}t + a_{21}x + a_{22}y + a_{23}z \,,$$
$$z' = a_{30}t + a_{31}x + a_{32}y + a_{33}z \,. \tag{6.1}$$

To determine the coefficients a_{ik}, we cast these equations into the following form:

$$t' = A\,t + \mathbf{B} \cdot \mathbf{r},$$
$$\mathbf{r}' = C\,\mathbf{r} + (\mathbf{D} \cdot \mathbf{r})\,\mathbf{w} + \mathbf{E}\,t. \tag{6.2}$$

The real number A is the same as a_{00}, the vector $\mathbf{B}$ has the components (a_{01}, a_{02}, a_{03}), and the vector $\mathbf{E}$ has the components (a_{10}, a_{20}, a_{30}). The part of $\mathbf{r}'$ that is linear in $\mathbf{r}$ must be constructed out of the only available vectors, namely $\mathbf{w}$ and $\mathbf{r}$ itself. It therefore has two components, one parallel to $\mathbf{r}$ and one parallel to $\mathbf{w}$. Because the component parallel to $\mathbf{w}$ must also be linear in $\mathbf{r}$, its coefficient must be a scalar product formed out of $\mathbf{r}$ and another vector $\mathbf{D}$.

Our next order of business is to find the real numbers A and C and the vectors $\mathbf{B}$, $\mathbf{D}$, and $\mathbf{E}$. They can only depend on $\mathbf{w}$. More specifically, A and C must be functions of the magnitude w of $\mathbf{w}$, and $\mathbf{B}$, $\mathbf{D}$, and $\mathbf{E}$ must be parallel to $\mathbf{w}$ with magnitudes depending solely on w. Thus,

$$t' = a(w)\,t + b(w)\,\mathbf{w} \cdot \mathbf{r}, \tag{6.3}$$

$$\mathbf{r}' = c(w)\,\mathbf{r} + \frac{d(w)}{w^2}(\mathbf{w} \cdot \mathbf{r})\,\mathbf{w} + e(w)\,\mathbf{w}\,t. \tag{6.4}$$

Problem 6.1. *What is achieved by the division by w^2 in the second term on the right-hand side of Eq. (6.4)?*

We obtain our first constraint on the five functions of w by considering an object $\mathcal{O}$ whose position relative to $\mathcal{F}_1$ is given by $\mathbf{r} = \mathbf{w}t$, so that

$$\mathbf{r}' = [c(w) + d(w) + e(w)]\,\mathbf{w}\,t.$$

Because both $\mathcal{O}$ and $\mathcal{F}_2$ move with velocity $\mathbf{w}$ relative to $\mathcal{F}_1$, $\mathcal{O}$ is actually at rest in $\mathcal{F}_2$. Hence

$$c(w) + d(w) + e(w) = 0. \tag{6.5}$$

To obtain further constraints, we make use of the inverse transformation. Since $\mathcal{F}_1$ moves with velocity $-\mathbf{w}$ relative to $\mathcal{F}_2$, this is given by

$$t = a(w)\,t' - b(w)\,\mathbf{w} \cdot \mathbf{r}', \tag{6.6}$$

$$\mathbf{r} = c(w)\,\mathbf{r}' + \frac{d(w)}{w^2}(\mathbf{w} \cdot \mathbf{r}')\,\mathbf{w} - e(w)\,\mathbf{w}\,t'. \tag{6.7}$$

We now simplify matters further by choosing the space axes so that $\mathbf{w}$ has the components $(w, 0, 0)$. Equations (6.3) and (6.4) then reduce to

$$t' = at + bwx, \quad x' = cx + dx + ewt, \quad y' = cy, \quad z' = cz, \tag{6.8}$$

while Eqs. (6.6) and (6.7) reduce to

$$t = at' - bwx', \quad x = cx' + dx' - ewt', \quad y = cy', \quad z = cz'. \tag{6.9}$$

Plugging Eqs. (6.8) into Eqs. (6.9), we obtain

$$
\begin{aligned}
t &= a\,(at + bwx) - bw\,(cx + dx + ewt) \\
&= (a^2 - bew^2)\,t + (abw - bcw - bdw)\,x\,, \tag{6.10}
\end{aligned}
$$

$$
\begin{aligned}
x &= c\,(cx + dx + ewt) + d\,(cx + dx + ewt) - ew\,(at + bwx) \\
&= (c^2 + 2cd + d^2 - bew^2)\,x + (cew + dew - aew)\,t\,, \tag{6.11}
\end{aligned}
$$

$$y = c^2 y\,, \tag{6.12}$$

$$z = c^2 z\,. \tag{6.13}$$

To satisfy Eq. (6.10), we must have that

$$a^2 - bew^2 = 1\,, \tag{6.14}$$

$$abw - bcw - bdw = 0\,, \tag{6.15}$$

and to satisfy Eq. (6.11), we must have that

$$c^2 + 2\,cd + d^2 - bew^2 = (c + d)^2 - bew^2 = 1\,, \tag{6.16}$$

$$cew + dew - aew = 0\,. \tag{6.17}$$

Using Eq. (6.5) and the fact that $w \neq 0$, the last four equations reduce to

$$a^2 - bew^2 = 1\,, \tag{6.18}$$

$$b(a + e) = 0\,, \tag{6.19}$$

$$e^2 - bew^2 = 1\,, \tag{6.20}$$

$$e(e + a) = 0\,. \tag{6.21}$$

Equation (6.20) implies that $e \neq 0$. With this, Eq. (6.21) tells us that $e = -a$. Equation (6.18) then implies that $b = (1 - a^2)/aw^2$. From Eq. (6.12) or (6.13) we learn that $c^2 = 1$, but since we have assumed that the space axes of the two reference frames are parallel (rather than antiparallel), we conclude that $c = 1$. Invoking once more Eq. (6.5), we find that $d = a - 1$. With a single unknown function left, Eq. (6.8) reduces to

$$t' = a(w)\,t + \frac{1 - a^2(w)}{a(w)\,w}\,x\,, \tag{6.22}$$

$$x' = a(w)\,x - a(w)\,wt\,, \quad y' = y\,, \quad z' = z\,. \tag{6.23}$$

A further constraint can be obtained by combining two such transformations. If $\mathcal{F}_3$ moves with the speed v relative to $\mathcal{F}_2$, the transformation from $\mathcal{F}_2$ to $\mathcal{F}_3$ is given by

$$t'' = a(v)\, t' + \frac{1 - a^2(v)}{a(v)\, v}\, x', \tag{6.24}$$

$$x'' = a(v)\, x' - a(v)\, vt', \quad y'' = y', \quad z'' = z'. \tag{6.25}$$

Plugging (6.24) and (6.25) into (6.22) and (6.23) yields

$$t'' = a(v) \left[a(w)\, t + \frac{1 - a^2(w)}{a(w)\, w}\, x \right] + \frac{1 - a^2(v)}{a(v)\, v} \left[a(w)\, x - a(w)\, wt \right]$$

$$= \underbrace{\left[a(v)\, a(w) - \frac{1 - a^2(v)}{a(v)\, v}\, a(w)\, w \right]}_{\star} t + [\dots]\, x, \tag{6.26}$$

$$x'' = a(v) \left[a(w)\, x - a(w)\, wt \right] - a(v)\, v \left[a(w)\, t + \frac{1 - a^2(w)}{a(w)\, w}\, x \right]$$

$$= \underbrace{\left[a(v)\, a(w) - a(v)\, v\, \frac{1 - a^2(w)}{a(w)\, w} \right]}_{\star} x - [\dots]\, t. \tag{6.27}$$

Another way of writing (6.26) and (6.27) is

$$t'' = \underbrace{a(u)}_{\star}\, t + \frac{1 - a^2(u)}{a(u)\, u}\, x, \tag{6.28}$$

$$x'' = \underbrace{a(u)}_{\star}\, x - a(u)\, ut, \quad y' = y, \quad z' = z, \tag{6.29}$$

where u is the speed of $\mathcal{F}_3$ relative to $\mathcal{F}_1$. The expressions marked by a star must be equal; hence

$$a(u) = a(v)\, a(w) - \frac{1 - a^2(v)}{a(v)\, v}\, a(w)\, w = a(v)\, a(w) - a(v)\, v\, \frac{1 - a^2(w)}{a(w)\, w}, \tag{6.30}$$

and thus

$$K \overset{\text{Def}}{=} \frac{1 - a^2(w)}{a^2(w)\, w^2} = \frac{1 - a^2(v)}{a^2(v)\, v^2}. \tag{6.31}$$

Since this holds for arbitrary w and v, K is a universal constant. Solving the first equality for $a(w)$, we arrive at

$$a(w) = \frac{1}{\sqrt{1 + Kw^2}}. \tag{6.32}$$

With this, the transformation (6.8) takes the form

$$t' = \frac{t + Kwx}{\sqrt{1 + Kw^2}}, \quad x' = \frac{x - wt}{\sqrt{1 + Kw^2}}, \quad y' = y, \quad z' = z. \tag{6.33}$$

Trumpets, please! We managed to reduce five unknown functions to a single unknown constant. What is more, if $K \neq 0$, its value depends on conventional units—those of an inverse velocity squared. We are therefore left with three physically distinct options:

   (i) $K = 0$, (ii) $K > 0$, (iii) $K < 0$.

6.3 Composition of velocities

$K = 0$ yields the *Galilean transformations* of Newtonian mechanics:

$$t' = t, \quad x' = x - wt. \tag{6.34}$$

Problem 6.2. *In this case* $u = v + w$.

If $K \neq 0$, we can use Eqs. (6.30) and (6.31) to obtain

$$a(u) = a(v)\, a(w)(1 - Kvw). \tag{6.35}$$

Problem 6.3. (*) *In this case*

$$u = \frac{v + w}{1 - Kvw}. \tag{6.36}$$

If $K > 0$, we can introduce another constant $\tilde{c} = 1/\sqrt{K}$, which has the dimension of a velocity, and write

$$u = \frac{v + w}{1 - vw/\tilde{c}^2}. \tag{6.37}$$

Setting $v = w = \tilde{c}/2$, we obtain $u = (4/3)\,\tilde{c}$: the speed of $\mathcal{F}_3$ relative to $\mathcal{F}_1$ is greater than the sum of the speeds of $\mathcal{F}_3$ relative to $\mathcal{F}_2$ and of $\mathcal{F}_2$ relative to $\mathcal{F}_1$. As both v and w approach $\tilde{c}$, u approaches infinity. And if the product vw is greater than $\tilde{c}^2$, u is negative!

Problem 6.4. *Verify these statements.*

If $K < 0$, we can introduce the constant $c = 1/\sqrt{-K}$, which again has the dimension of a velocity, and write

$$u = \frac{v + w}{1 + vw/c^2}. \tag{6.38}$$

Setting $v = w = c/2$, we obtain $u = (4/5)\,c$: the speed of $\mathcal{F}_3$ relative to $\mathcal{F}_1$ is less than the sum of the speeds of $\mathcal{F}_3$ relative to $\mathcal{F}_2$ and of $\mathcal{F}_2$ relative to $\mathcal{F}_1$. And if v or w approaches c, then so does u.

Problem 6.5. *Verify these statements.*

6.4   The case against positive K

Pick an infinitesimal segment $d\mathcal{C}$ of a spacetime path $\mathcal{C}$. In terms of $\mathcal{F}_1$, it has the components (dt, dx, dy, dz), and in terms of $\mathcal{F}_2$, it has the components (dt', dx', dy', dz'). These components, too, are related to each other via the transformation (6.33):

$$dt' = \frac{dt + Kw\,dx}{\sqrt{1 + Kw^2}}, \quad dx' = \frac{dx - w\,dt}{\sqrt{1 + Kw^2}}, \quad dy' = dy, \quad dz' = dz. \quad (6.39)$$

Problem 6.6.

$$dt'^2 + K\left(dx'^2 + dy'^2 + dz'^2\right) = dt^2 + K\left(dx^2 + dy^2 + dz^2\right). \quad (6.40)$$

If $K = 0$, this doesn't tell us anything new.

If $K > 0$, it is convenient to choose measurement units in which $K = 1$, so that distances and durations are measured in the same units. What you have just proved is that the expression

$$d\tilde{s}^2 \overset{\text{Def}}{=} dt^2 + dx^2 + dy^2 + dz^2 \quad (6.41)$$

is invariant under the transformation (6.39) with K equal to unity.

This should remind you of something. If x, y, z are the components of a vector $\mathbf{r}$, then the expression $x^2 + y^2 + z^2$ is invariant under coordinate transformations that preserve the magnitudes of vectors. Here we have an analogous expression that is invariant under the transformations (6.39). If $K > 0$ were Nature's choice, we could identify $d\tilde{s}$ as the "length" of the path segment $d\mathcal{C}$, which we are looking for, as you will recall from (Sec. 5.7).

But is $K = 1$ a live option? Consult Fig. 6.1. The speed of $\mathcal{F}_2$ relative to $\mathcal{F}_1$ is $w = \tan\alpha$. The coordinates of a spacetime vector $\mathbf{v}$ in the x–t plane are therefore related via

$$t' = \frac{t + x\tan\alpha}{\sqrt{1 + \tan^2\alpha}}, \quad x' = \frac{x - t\tan\alpha}{\sqrt{1 + \tan^2\alpha}}. \quad (6.42)$$

Problem 6.7. $(*)$

$$t' = t\cos\alpha + x\sin\alpha, \quad x' = x\cos\alpha - t\sin\alpha. \quad (6.43)$$

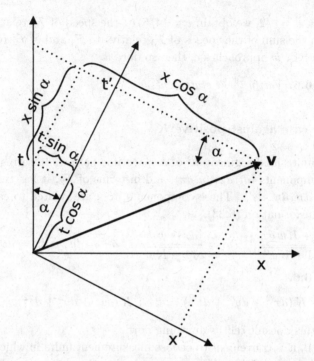

Fig. 6.1   Effect of a clockwise rotation of the x and t axes (Eq. 6.43) on the components of a vector in a plane containing these axes.

This describes the effect of a clockwise rotation of the x and t axes on the components of a vector in a plane containing these axes. Now we know why u approaches infinity as both v and w approach $\tilde{c}$. (If K equals 1, so does $\tilde{c}$.) The composition of two rotations of the x–t plane by $45°$ yields a rotation by $90°$. This turns the t axis into the x axis. If $w = v = \tilde{c}$, then an object at rest in $\mathcal{F}_3$ moves along the x axis of $\mathcal{F}_1$; for an observer at rest in $\mathcal{F}_1$, it therefore moves with an infinite speed.

We also see why the composition of two positive speeds can yield a negative speed. The reason why u comes out negative if $vw > \tilde{c}^2$ is not that an object at rest in $\mathcal{F}_3$ moves in the direction of the negative x axis but that, for an observer at rest in $\mathcal{F}_1$, it moves backward in time.

But if making a U–turn is as easy in a spacetime plane containing the time axis as it is in a plane containing two space axes, no coherent physics can result. If what is a space axis for one observer can be the time axis for another, then the difference between space and time depends on the language we use to describe a physical situation, rather than on the physical situation itself. Yet any viable theory about the physical world must, at a minimum, feature an objective—i.e., language-independent—difference between space and time.

Here is one reason why. The stability of an atom—and with it that of matter—rests crucially on the fuzziness of its internal relative positions (Sec. 4.3). At the same time the position of an atomic electron relative to the nucleus cannot be *completely* fuzzy, for in this case the probability per unit volume of detecting it would be constant over the (abstract) space of sharp positions relative to the nucleus. In other words, the electron's position probability distribution (relative to the nucleus) would be homogeneous. But this cannot be: the electron has to be more likely to be detected in one place rather than in another. On the other hand, there has to be a ground state that is homogeneous in time: the probabilities it defines must not change with time. These two requirements cannot be met unless there is a language-independent difference between space and time.

6.5 An invariant speed

If $K \leq 0$, there exists an invariant speed. (Whatever moves with an *invariant speed* relative to one inertial frame, moves with the same speed relative to any other inertial frame.)

To see this, suppose that an object traveling with c relative to $\mathcal{F}_2$ also travels with c relative to $\mathcal{F}_3$, regardless of the relative speed v between $\mathcal{F}_2$ and $\mathcal{F}_3$. Inserting $w = c$ and $u = c$ into Eq. (6.36), we obtain $c = 1/\sqrt{-K}$. Thus an invariant speed equal to c exists if $K \leq 0$ but not if $K > 0$.

If $K - 0$, the invariant speed is infinite: what travels with infinite speed relative to one inertial frame, travels with infinite speed relative to every inertial frame. In other words, if the departure of an object and its arrival are simultaneous in one inertial frame, they are simultaneous in every inertial frame. In Newtonian mechanics, simultaneity is therefore absolute: whether or not two events happen at the same time is independent of the reference frame that we use to describe their spatiotemporal relations.

If K is negative, the invariant speed is finite. Empirically we know that an invariant speed exists, and that it is finite. We call it the "speed of light (in vacuum)." Nature's choice—$K < 0$—stands thereby revealed.

The invariant speed—be it infinite or finite—cannot be attained by any object that can be at rest. To see this, imagine that you spend a finite amount of fuel accelerating from zero to some speed $v < c$. In the frame in which you are now at rest, the difference between the invariant speed and yours is still the invariant speed. You may repeat the procedure as many times as you wish—the difference between the invariant speed and yours remains the invariant speed. No finite amount of fuel is sufficient to reach it. Moreover, since you have to attain the invariant speed before you can make a U-turn in a plane that contains the time axis, the existence of an invariant speed is what prevents you from going back in time.

6.6   Proper time

With $K = -1/c^2$, the transformation (6.39) and the invariant expression (6.40) take the following forms:

$$t' = \frac{t - wx/c^2}{\sqrt{1 - w^2/c^2}}, \quad x' = \frac{x - wt}{\sqrt{1 - w^2/c^2}}, \quad y' = y, \quad z' = z, \qquad (6.44)$$

$$ds^2 = dt^2 - (dx^2 + dy^2 + dz^2)/c^2. \qquad (6.45)$$

Here at last is the wanted "length" ds of a path segment $d\mathcal{C}$ with components dt, dx, dy, dz (Sec. 5.7). What is its physical significance?

Imagine a clock that travels along the infinitesimal segment $d\mathcal{C}$ of a spacetime path $\mathcal{C}$.[2] The passing of time it shows is the passing of the coordinate time of the inertial frame in which it is *momentarily* (i.e., during an infinitesimal time interval) at rest. To distinguish this time from the coordinate time of a fixed inertial frame, we refer to it as *proper time*.[3] The proper time $s[\mathcal{C}]$ associated with an entire spacetime path is thus related to a system of inertial coordinates via

$$\int_{\mathcal{C}} ds = \int_{\mathcal{C}} \sqrt{dt^2 - (dx^2 + dy^2 + dz^2)/c^2} = \int_{\mathcal{C}} dt \sqrt{1 - v^2/c^2}. \qquad (6.46)$$

---

[2] A word of caution. Because we cannot help visualizing a plane containing the time axis as a spatial plane, like a figure on a page, we are tempted to think about the spatiotemporal whole as if it were a 4-dimensional spatial landscape, and about the present—a 3-dimensional section of this landscape—as if it advanced through this spatiotemporal whole. This way of thinking involves an illegitimate duplication of time. First we mentally represent time as a dimension of a 4-dimensional whole, and then we think of this 4-dimensional whole (including its temporal dimension) as a spatial whole that persists in *another* temporal dimension, in which the present advances through this now spatially conceived whole. So if we say that an object "travels" along a spacetime path, we must take care not to think of this object as traveling along a path in a *persistent* 4-dimensional expanse.

[3] Because we only have $ds^2 = dt^2$ (in the momentary rest frame), there are two possibilities: $ds = dt$ and $ds = -dt$. As we shall see in Sec. 15.2, the former holds for particles while the latter holds for *antiparticles* [Costella *et al.* (1997)].

The definition of proper time is restricted to timelike paths. A finite path is called *timelike* if every one of its segments is, and an infinitesimal path segment $d\mathcal{C}$ is timelike if $ds^2 > 0$. By the same token, a finite path is called *spacelike* if every one of its segments is, and an infinitesimal path segment $d\mathcal{C}$ is spacelike if $ds^2 < 0$. Finally, a finite path is called *lightlike* or *null* if every one of its segments is, and an infinitesimal path segment $d\mathcal{C}$ is lightlike or null if $ds^2 = 0$.

Problem 6.8. *The speed of an object moving along a timelike path is always less than the invariant speed c.*

Problem 6.9. (∗) *Use* $dt^2 - dx^2/c^2 = (dt')^2 - (dx')^2/c^2$ *to show that c is an invariant speed.*

6.7 The meaning of mass

Now we are ready to address the question left pending in Chap. 5, concerning the significance of the single particle-specific parameter b, which appears in the propagator for a free and stable particle,

$$\langle \mathbf{r}_B, t_B | \mathbf{r}_A, t_A \rangle = \int \mathcal{D}\mathcal{C} \, e^{ibs[\mathcal{C} | \mathbf{r}_A, t_A \to \mathbf{r}_B, t_B]} . \tag{6.47}$$

If proper time is measured in seconds, b is measured in radians per second. Let us follow a particle as it travels from A to B along $\mathcal{C}$. As its proper time s increases, the phase factor e^{ibs} rotates in the complex plane. We may therefore think of it as a *clock*. Each time a cycle is completed, it "ticks." Although a feature of the quantum-mechanical probability calculus rather than of the physical world, this clock reveals a deep connection between the quantum-mechanical probability calculus and the metric properties of the physical world.

It is customary—

- to insert a minus (so that our clock actually turns clockwise!): e^{-ibs},
- to multiply by 2π (so that we may think of b as the rate at which the clock "ticks"—the number of cycles it completes each second): $e^{-i2\pi bs}$,
- to divide by Planck's constant h (so that b is measured in energy units, in which case it is known as the particle's *rest energy*): $e^{-i(2\pi/h)bs} = e^{-(i/\hbar)bs}$,
- to multiply by c^2 (so that b is measured in mass units, in which case it is known as the particle's *mass m*): $e^{-(i/\hbar)bc^2 s} = e^{-(i/\hbar)mc^2 s}$.

If we use natural units, in which $\hbar = c = 1$, we are back at e^{-ibs} with $b = m$. Thus apart from a factor 2π, which converts between cycles per second and radians per second, the mass of a particle is the rate at which the clock associated with the particle ticks.

Now imagine two ordinary clocks initially at rest relative to the same inertial frame $(\mathcal{F}_1)$. They are located in about the same place and they tick at the same rate. They then travel along different spacetime paths. Eventually they are again at rest relative to the same inertial frame $(\mathcal{F}_2)$ and located in about the same place. Will they still tick at the same rate? Indeed they will, but why? This is not something we should take for granted. The reason why they still tick at the same rate is that (i) the rates at which clocks tick depend on the rates at which free particles tick in their rest frames, (ii) the rate at which a free particle ticks in its rest frame is determined by its mass, and (iii) the mass of a free particle is independent of the particle's location in spacetime.

6.8   The case against $K = 0$

If K were to vanish, proper time would be the same as coordinate time. Every spacetime path leading from A to B would contribute the same amplitude $e^{ib(t_B - t_A)}$ to the propagator (6.47), which would be hopelessly divergent as a result—as well as independent of the distance between A and B. To obtain finite probabilities, cancellations ("destructive interference") must occur. The phase factors $e^{ibs[C|\mathbf{r}_A,t_A \to \mathbf{r}_B,t_B]}$ must not all point in the same direction in the complex plane.

While this does not quite rule out the option $K = 0$ (after all, there is such a thing as non-relativistic quantum mechanics), the manner in which one obtains the non-relativistic substitute for ds makes it clear that this can only be an approximation—useful to the extent that all powers of v^2/c^2 but the first can be ignored.

Problem 6.10. *Use the Taylor series to show that*

$$\sqrt{1 - \frac{v^2}{c^2}} = 1 - \frac{1}{2}\frac{v^2}{c^2} - \frac{1}{8}\frac{v^4}{c^4} - \cdots \approx 1 - \frac{1}{2}\frac{v^2}{c^2}. \tag{6.48}$$

In this approximation, Eq. (6.46) becomes

$$\int_C ds \approx \int_C dt \left(1 - \frac{1}{2}\frac{v^2}{c^2}\right) = (t_B - t_A) - \frac{1}{2c^2}\int_C dt\, v^2, \tag{6.49}$$

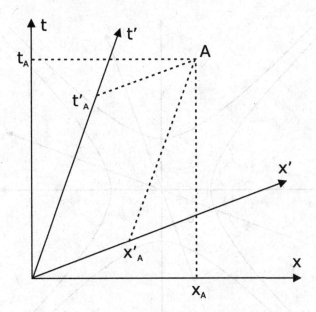

Fig. 6.2 Spacetime coordinates of an event A relative to two different inertial frames.

and the propagator (6.47) takes the form

$$\langle \mathbf{r}_B, t_B | \mathbf{r}_A, t_A \rangle = \int \mathcal{D} \mathcal{C} \, e^{-(i/\hbar)mc^2 \int_c ds} \tag{6.50}$$

$$\approx e^{-(i/\hbar)mc^2(t_B - t_A)} \int \mathcal{D} \mathcal{C} \, e^{(i/\hbar) \int_c dt \, (mv^2/2)} . \tag{6.51}$$

Since the phase factor $e^{-(i/\hbar)mc^2(t_B - t_A)}$ is the same for all paths from A to B, it has no effect on the probabilities that the propagator serves to calculate. Hence the following may be used as the *non-relativistic* propagator:

$$\langle \mathbf{r}_B, t_B | \mathbf{r}_A, t_A \rangle = \int \mathcal{D} \mathcal{C} \, e^{(i/\hbar) \int_c dt \, (mv^2/2)} . \tag{6.52}$$

6.9 Lorentz transformations: Some implications

Let us explore, briefly, what is implied by the Lorentz transformation (6.44). We shall use units in which $c = 1$.

Setting $t' = 0$, we obtain $t = wx$. This tells us that the slope of the x' axis in the x–t frame $(\mathcal{F}_1)$ equals w. Setting $x' = 0$, we obtain $t = x/w$.

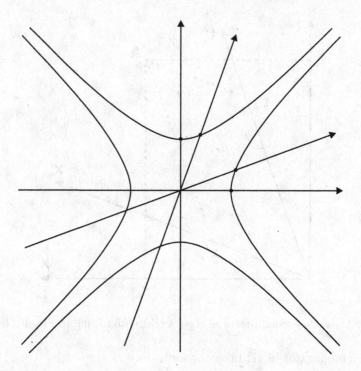

Fig. 6.3   The unit points of the space and time axes are situated on hyperbolas.

This tells us that the slope of the t' axis in the x–t frame is the inverse: $1/w$. The axes of $\mathcal{F}_2$ are thus rotated relative to $\mathcal{F}_1$ by equal angles but in *opposite* directions (Fig. 6.2). The spacetime path of a (classical) light signal traveling in the direction defined by the x axis is therefore the bisector of the angle between the x and t axes of *both* frames.

The big news here is that simultaneity is relative. Events that happen at the same time in a given frame can be connected by a line parallel to the frame's space axis, just as events that happen in the same place in a given frame can be connected by a line parallel to the frame's time axis. This is true in any case. But since in this case the space axes of the two frames are not parallel, whether or not two events are simultaneous depends on the language—the inertial reference frame—that we use to describe a physical situation, rather than on the physical situation itself. If two events E_1, E_2 are simultaneous relative to one frame, there are frames relative to which E_1 happens after E_2 as well as frames relative to which E_1 happens before E_2.

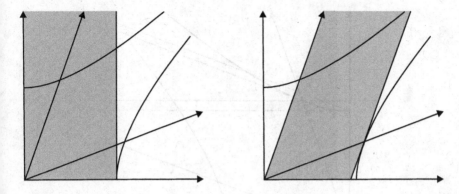

Fig. 6.4   Lorentz contraction.

Next we want to determine the unit points on the space and time axes of the two inertial frames of Fig. 6.2. To this end we use the invariant expression

$$t^2 - x^2 = t'^2 - x'^2 .$$

The unit point of the time axis of $\mathcal{F}_2$ has the coordinates $t' = 1, x' = 0$; it therefore lies on the hyperbola defined by $t^2 - x^2 = 1$. The unit point of the space axis of $\mathcal{F}_2$ has the coordinates $t' = 0, x' = 1$; it therefore lies on the hyperbola defined by $x^2 - t^2 = 1$ (Fig. 6.3).

Figure 6.4 illustrates the phenomenon of *Lorentz contraction*: a moving object is shortened in the direction in which it is moving. The graph on the left shows the spacetime area traced by a meter stick at rest in $\mathcal{F}_1$. In $\mathcal{F}_2$, in which the stick is moving, it is shorter: the distance between its end points *on the x' axis* is less than a meter. The graph on the right shows the spacetime area traced by a meter stick at rest in $\mathcal{F}_2$. In $\mathcal{F}_1$, in which this stick is moving, it is shorter: the distance between its end points *on the x axis* is again less than a meter.

Figure 6.5 illustrates the phenomenon of *time dilation*: a moving clock runs slower than a clock at rest. Let clock 1 be at rest in $\mathcal{F}_1$, its spacetime path being the t axis. Let clock 2 be at rest in $\mathcal{F}_2$, its spacetime path being the t' axis. In the language of $\mathcal{F}_1$, when a second has passed (on clock 1), less than a second has passed on clock 2. In the language of $\mathcal{F}_2$, when a second has passed (on clock 2), less than a second has passed on clock 1.

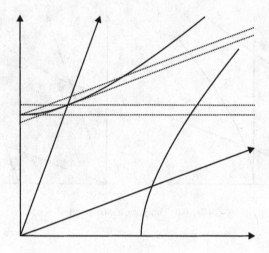

Fig. 6.5   Time dilation.

6.10   4-vectors

For later use we still need to define 4-vectors and their scalar product. A *4-vector* is a quadruplet of real numbers that changes like the coordinates of a point in spacetime when subjected to a Lorentz transformation. The *scalar product* of the 4-vectors $\vec{a} = (a_t, \mathbf{a}) = (a^0, a^1, a^2, a^3)$ and $\vec{b} = (b_t, \mathbf{b}) = (b^0, b^1, b^2, b^3)$ is

$$(\vec{a}, \vec{b}) \overset{\text{Def}}{=} a^0 b^0 - a^1 b^1 - a^2 b^2 - a^3 b^3 . \tag{6.53}$$

To check that $(\vec{a}, \vec{b})$ is invariant under Lorentz transformations, we consider the sum of two 4-vectors, $\vec{c} = \vec{a} + \vec{b}$, and calculate

$$(\vec{c}, \vec{c}) = (\vec{a} + \vec{b}, \vec{a} + \vec{b}) = (\vec{a}, \vec{a}) + (\vec{b}, \vec{b}) + 2(\vec{a}, \vec{b}) . \tag{6.54}$$

Because the expression (6.45) is invariant under the transformation (6.44), the "squares" $(\vec{a}, \vec{a})$, $(\vec{b}, \vec{b})$ and $(\vec{c}, \vec{c})$ are so, too ($c = 1$ was assumed). But if these are invariant, then so is the last term on the right-hand side of Eq. (6.54). In other words, $(\vec{a}, \vec{b})$ is a *4-scalar*.

Chapter 7

The Feynman route to Schrödinger (stage 2)

7.1 Action

Let us return to Eq. (5.9). For an infinitesimal path made up of two segments, this takes the form

$$Z(d\mathcal{C}) = Z(d\mathcal{C}_2) Z(d\mathcal{C}_1). \tag{7.1}$$

In Sec. 5.6 we noted that, in the general case, $Z(d\mathcal{C})$ is a function of the spacetime position of $d\mathcal{C}$ and of the coordinate differentials $(dt, d\mathbf{r})$. Equation (7.1) is therefore tantamount to

$$Z(t, \mathbf{r}, dt_1 + dt_2, d\mathbf{r}_1 + d\mathbf{r}_2) = Z(t, \mathbf{r}, dt_2, d\mathbf{r}_2) Z(t, \mathbf{r}, dt_1, d\mathbf{r}_1). \tag{7.2}$$

From this we conclude—by the same reasoning we used in Sec. 5.6—that the amplitude associated with an infinitesimal path segment takes the form

$$Z(t, \mathbf{r}, dt, d\mathbf{r}) = e^{z\, dS(t,\mathbf{r},dt,d\mathbf{r})}. \tag{7.3}$$

We further conclude—by the same reasoning we used in Sec. 5.7—that for a *stable* particle the complex number $z = a + ib$ has to be imaginary ($a = 0$). It is customary to set $b = 1/\hbar$, so that the infinitesimal *action* dS is measured in conventional units (energy × time or momentum × distance):

$$Z(t, \mathbf{r}, dt, d\mathbf{r}) = e^{(i/\hbar)dS(t,\mathbf{r},dt,d\mathbf{r})}. \tag{7.4}$$

A glance at Eq. (6.50) now reveals the action differential for a free and stable particle:

$$dS = -mc^2 ds. \tag{7.5}$$

Equation (7.2) can be generalized to read

$$Z(t, \mathbf{r}, u\, dt, u\, d\mathbf{r}) = [Z(t, \mathbf{r}, dt, d\mathbf{r})]^u. \tag{7.6}$$

69

In conjunction with Eq. (7.4) this tells us that

$$dS(t, \mathbf{r}, u\,dt, u\,d\mathbf{r}) = u\,dS(t, \mathbf{r}, dt, d\mathbf{r}). \tag{7.7}$$

Technically this makes dS *homogeneous* (of first degree) in the differentials dt and $d\mathbf{r}$. This homogeneity allows us to interpret dS as defining a *differential geometry*—the kind of geometry one needs for assigning lengths to paths on a warped surface, in a warped space, or in a warped spacetime.

7.2   How to influence a stable particle?

Suppose now that something—no matter what—exerts an influence—no matter how—on a stable particle. How do we incorporate effects on the motion of such a particle? Since all we have at our disposal is the lengths of paths as defined by this differential geometry, the only way in which we can formulate effects on the particle's motion is through modifications of dS. Since the action differential encapsulates the particle's observable behavior, it is moreover essential that it be invariant under Lorentz transformations. It must not depend on the language—the reference frame—that we use to describe the particle's behavior. The scope of possible effects on the motion of a particle is therefore limited to modifications of the action differential (7.5) that preserve

(1) the homogeneity expressed by Eq. (7.7),
(2) the invariance of dS as a 4-scalar.

One possible modification that meets these requirements stands out as the most obvious and straightforward: to preserve Lorentz invariance, we add to (7.5) the scalar product (6.53) of two 4-vectors, and to preserve homogeneity, we let one of these vectors be $d\vec{r} = (c\,dt, d\mathbf{r})$. We shall denote the other vector, whose components are in general functions of t and $\mathbf{r}$, by $\vec{A} = (V, \mathbf{A})$. The inclusion of a charge q allows the extent to which particles are affected to differ between particle species. Thus:

$$dS = -mc^2\,ds - qV(t, \mathbf{r})\,dt + \frac{q}{c}\mathbf{A}(t, \mathbf{r}) \cdot d\mathbf{r}. \tag{7.8}$$

Here Gaussian units are used. In SI units the factor $1/c$ is absorbed into the definition of $\mathbf{A}$:

$$dS = -mc^2\,ds - qV(t, \mathbf{r})\,dt + q\mathbf{A}(t, \mathbf{r}) \cdot d\mathbf{r}. \tag{7.9}$$

As we shall see in Sec. 9.5, the *fields* V and $\mathbf{A}$ uniquely determine the classical *electric field* $\mathbf{E}$ and the classical *magnetic field* $\mathbf{B}$. In other words, the action differential (7.8) encapsulates the effects of the electromagnetic force.[1]

7.3 Enter the wave function

In the non-relativistic approximation, the action differential (7.5) for a free particle takes the form

$$dS = \frac{m}{2}v^2\,dt\,, \tag{7.10}$$

as we gather from Eq. (6.52). If the vector potential vanishes or is small enough to be ignored, and if we are dealing with a particle of unit charge ($q = 1$), then the action differential (7.8) takes the simple form

$$dS = \left(\frac{m}{2}v^2 - V\right)dt\,. \tag{7.11}$$

We now introduce a function $\psi(t, \mathbf{r})$ such that

$$\psi(\mathbf{r}_B, t_B) = \int d^3r_A\,\langle\mathbf{r}_B, t_B|\mathbf{r}_A, t_A\rangle\,\psi(\mathbf{r}_A, t_A)\,. \tag{7.12}$$

Using the path–integral form of the propagator $\langle\mathbf{r}_B, t_B|\mathbf{r}_A, t_A\rangle$,

$$\int \mathcal{DC}\,e^{(i/\hbar)\int_{\mathcal{C}} dS}\,, \tag{7.13}$$

with the action differential (7.11), we arrive at

$$\psi(\mathbf{r}_B, t_B) = \int d^3r_A\left[\int \mathcal{DC}\,e^{(i/\hbar)\int_{\mathcal{C}}[(m/2)v^2 - V]dt}\right]\psi(\mathbf{r}_A, t_A)\,. \tag{7.14}$$

7.4 The Schrödinger equation

We obtain a more user-friendly expression to work with by considering an infinitesimal time interval $t_B - t_A = \epsilon$. In this case there is just one (infinitesimal) path connecting $(t_A, \mathbf{r}_A)$ and $(t_B, \mathbf{r}_B)$, and all that remains of the path integral is an as yet unknown normalization factor $\mathcal{N}$:

$$\psi(\mathbf{r}_B, t + \epsilon) = \mathcal{N}\int d^3r_A\,e^{(i/\hbar)[(m/2)v^2 - V]\epsilon}\,\psi(\mathbf{r}_A, t)\,. \tag{7.15}$$

---

[1]Strictly speaking, Eq. (7.8) encapsulates the possible electromagnetic effects on the observable behavior of a *scalar particle*. By this we mean a particle whose wave function has a single component. This excludes particles with spin (Chap. 12) and particles that are affected by nuclear forces (Sec. 15.9).

To further simplify our lives, we return to one spatial dimension,

$$\psi(\dot{x}_B, t + \epsilon) = \mathcal{N} \int_{-\infty}^{+\infty} dx_A \, e^{(i/\hbar)[(m/2)((x_b - x_A)^2/\epsilon^2) - V]\epsilon} \, \psi(x_A, t), \quad (7.16)$$

and we substitute η for $x_A - x_B$ (and drop the now superfluous index B):

$$\psi(x, t + \epsilon) = \mathcal{N} \int_{-\infty}^{+\infty} d\eta \, e^{(i/\hbar)[(m/2)(\eta^2/\epsilon^2) - V]\epsilon} \, \psi(x + \eta, t). \quad (7.17)$$

Take a look at the exponential factor $e^{im\eta^2/2\hbar\epsilon}$. In the limit $\epsilon \to 0$, this completes an infinite number of cycles as η increases by a finite amount unless η^2 is of the same order of magnitude as ϵ. Unless this is the case, the remaining factors of the integrand—$e^{-i\epsilon V/\hbar}$ and $\psi(x+\eta, t)$—can be treated as constant during a cycle, so that the net contribution to the integral from such a cycle is zero. Hence in order to take the limit $\epsilon \to 0$, we need to expand $e^{-i\epsilon V/\hbar} \psi(x + \eta, t)$ to the first power of ϵ or to the second power of η:

$$e^{-i\epsilon V/\hbar} \psi(x + \eta, t) \approx \left(1 - \frac{i\epsilon V}{\hbar}\right) \left(\psi(t, x) + \frac{\partial\psi}{\partial x}\eta + \frac{\partial^2\psi}{\partial x^2}\frac{\eta^2}{2}\right)$$

$$\approx \left(1 - \frac{i\epsilon V}{\hbar}\right) \psi(t, x) + \frac{\partial\psi}{\partial x}\eta + \frac{\partial^2\psi}{\partial x^2}\frac{\eta^2}{2}. \quad (7.18)$$

The following integrals need to be evaluated:

$$I_1 = \int d\eta \, e^{im\eta^2/2\hbar\epsilon}, \quad I_2 = \int d\eta \, e^{im\eta^2/2\hbar\epsilon}\eta, \quad I_3 = \int d\eta \, e^{im\eta^2/2\hbar\epsilon}\eta^2.$$

I_2 vanishes because the integrand is antisymmetric: the integral over the interval from $-\infty$ to 0 is the negative of the integral over the interval from 0 to $+\infty$. Using Eq. (3.28) we find that $I_1 = \sqrt{2\pi i\hbar\epsilon/m}$, and using Eq. (3.29) we find that $I_3 = \sqrt{2\pi\hbar^3\epsilon^3/im^3}$. Putting everything together, we arrive at

$$\psi(x, t) + \frac{\partial\psi}{\partial t}\epsilon = \mathcal{N}\sqrt{\frac{2\pi i\hbar\epsilon}{m}} \left(1 - \frac{i\epsilon}{\hbar}V(x, t)\right) \psi(x, t) + \frac{\mathcal{N}}{2}\sqrt{\frac{2\pi\hbar^3\epsilon^3}{im^3}} \frac{\partial^2\psi}{\partial x^2}. \quad (7.19)$$

For the two sides to be equal as ϵ tends to zero, we must have that $\mathcal{N} = \sqrt{m/2\pi i\hbar\epsilon}$. Equation (7.19) therefore reduces to

$$\frac{\partial\psi}{\partial t} = -\frac{i}{\hbar}V(x, t)\,\psi(x, t) + \frac{i\hbar}{2m}\frac{\partial^2\psi}{\partial x^2}. \quad (7.20)$$

Since ϵ has dropped out, all that remains to be done is multiply both sides with $i\hbar$. The result is the standard form of the Schrödinger equation for a particle of unit charge with one degree of freedom, moving under an influence represented by the potential $V(t,x)$ (cf. Eq. 4.4):

$$i\hbar\frac{\partial\psi}{\partial t} = -\frac{\hbar^2}{2m}\frac{\partial^2\psi}{\partial x^2} + V\psi. \tag{7.21}$$

Introducing the differential operator $\partial/\partial\mathbf{r}$—a machine that accepts a function $f(\mathbf{r})$ and returns a vector with the components $(\partial f/\partial x, \partial f/\partial y, \partial f/\partial z)$—we can write the corresponding equation for a particle of charge q with three degrees of freedom in the following way (cf. Eq. 4.5):

$$\left(i\hbar\frac{\partial}{\partial t} - qV\right)\psi = \frac{1}{2m}\left(\frac{\hbar}{i}\frac{\partial}{\partial\mathbf{r}}\right)\cdot\left(\frac{\hbar}{i}\frac{\partial}{\partial\mathbf{r}}\right)\psi. \tag{7.22}$$

As we shall see in Sec. 11.6, $i\hbar(\partial/\partial t)$ is the operator associated with the particle's total energy, and $(\hbar/i)(\partial/\partial\mathbf{r})$ is the operator associated with the particle's total momentum. (See Sec. 11.2 for the reason why operators are associated with observables.) On the left-hand side we thus have the operator for the particle's kinetic energy. The inclusion of the particle's potential momentum $(q/c)\,\mathbf{A}$ (see Eq. 9.20) is straightforward:

$$\left(i\hbar\frac{\partial}{\partial t} - qV\right)\psi = \frac{1}{2m}\left(\frac{\hbar}{i}\frac{\partial}{\partial\mathbf{r}} - \frac{q}{c}\mathbf{A}\right)\cdot\left(\frac{\hbar}{i}\frac{\partial}{\partial\mathbf{r}} - \frac{q}{c}\mathbf{A}\right)\psi. \tag{7.23}$$

PART 2
A Closer Look

Chapter 8

Why quantum mechanics?

8.1 The classical probability calculus

An exhaustive description of a classical physical system at any one time consists of the values of a fixed number N of *coordinates* and an equal number of *momenta*. Such a description is usually referred to as the system's *state*, and the system is said to have N *degrees of freedom*. This makes it possible to treat the state of a classical system with N degrees of freedom as a point $\mathcal{P}$ in a $2N$-dimensional space $\mathcal{S}$ known as *phase space*.

As an example consider the classical harmonic oscillator (Fig. 8.1). It has one degree of freedom, a 2-dimensional phase space, and the following equation of motion:

$$m\frac{d^2x}{dt^2} = -kx. \tag{8.1}$$

Problem 8.1. *In Fig. 8.1, which way does the oscillator's state $\mathcal{P}$ move? Clockwise or counterclockwise?*

Now for some truisms. A physical theory is as good as its predictions, and what quantum mechanics predicts are measurement outcomes. Moreover, with the exception of measurements that have finite sets of possible outcomes, no real-world experiment has an *exact* outcome. (Otherwise one could experimentally establish whether a physical quantity with a continuous range of possible values has a rational rather than irrational value.) If measurement outcomes are digitally displayed, as may reasonably be assumed, then every measurement has a finite number of possible outcomes. A (successful) measurement is therefore equivalent to a finite number of simultaneous measurements each having two possible outcomes. We will call such a measurement an *elementary test*. An elementary test associated

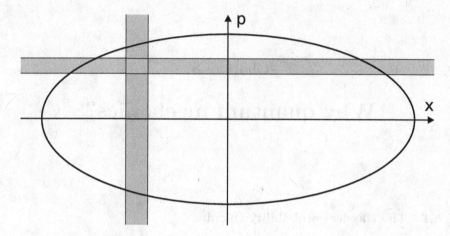

Fig. 8.1   Phase space diagram of a harmonic oscillator. Observe that when $|x|$ is at a maximum, p changes its sign (the oscillator reverses its motion), and when $x = 0$, $|p|$ is at a maximum (the oscillator moves fastest). See the main text for the significance of the gray strips.

with a continuous physical quantity X typically answers the question: does the value of X lie in the interval I? If the possible outcomes of a measurement are the intervals I_k, $k = 1, \ldots, n$, the outcome of one elementary test will be positive, while those of the remaining $n - 1$ tests will be negative.

In our oscillator example, intervals of the x axis and intervals of the p axis are some of the possible outcomes of elementary tests. In the system's phase space, the former intervals correspond to vertical strips, the latter to horizontal strips, as indicated in Fig. 8.1. The intersection of a horizontal and a vertical strip corresponds to the outcome of another elementary test, determining the simultaneous truth of the propositions "the value of x lies in the vertical strip" and "the value of p lies in the horizontal strip." Generalizing this possibility by allowing every *subset* of phase space to be the possible outcome of an elementary test, we arrive at the following characterization of the classical probability calculus.

- An elementary test is (represented by) a subset $\mathcal{U}$ of a phase space $\mathcal{S}$.
- The algorithm that serves to assign probabilities to the outcomes of elementary tests is a point $\mathcal{P}$ in $\mathcal{S}$.
- The probability of obtaining a positive outcome for $\mathcal{U}$ is 1 if $\mathcal{U}$ contains $\mathcal{P}$ (in set-theoretic notation, $\mathcal{P} \in \mathcal{U}$).
- The probability of obtaining a positive outcome for $\mathcal{U}$ is 0 if $\mathcal{U}$ does not contain $\mathcal{P}$ ($\mathcal{P} \notin \mathcal{U}$).

This probability algorithm is *trivial*, in the sense that it only assigns trivial probabilities: 0 or 1. Because it is trivial, $\mathcal{P}$ can be thought of as a *state* in the classical sense of the word: a collection of possessed properties. We are licensed to believe that $\mathcal{U}$ represents a physical property (rather than an elementary test), and that if the probability of finding this property is 1, it is *because* the system possesses the property, irrespective of whether the corresponding elementary test is made.

8.2 Why nontrivial probabilities?

One of the most spectacular failures of classical physics was its inability to account for the stability of matter. "Ordinary" material objects

- have spatial extent (they "occupy space"),
- are composed of a (large but) finite number of objects without spatial extent (particles that do not "occupy space"),
- and are stable (they neither explode nor collapse as soon as they are created).

Ordinary objects occupy as much space as they do because atoms and molecules occupy as much space as *they* do. So how is it that a hydrogen atom in its ground state occupies a space roughly one tenth of a nanometer across? Thanks to quantum mechanics, we now understand that the stability of ordinary objects rests on the indeterminacy or *fuzziness* of the relative positions and momenta of their constituents (Sec. 4.3).

What, then, is the proper (i.e., mathematically rigorous and philosophically sound) way to define and quantify a fuzzy property or value? It is to assign *nontrivial* probabilities—probabilities *between* 0 and 1—to the possible outcomes of a measurement of this property. To be precise, the proper way of quantifying a fuzzy property or value is to make *counterfactual* probability assignments. (To assign a probability to a possible measurement outcome counterfactually—contrary to the facts—is to assign it to a possible outcome of a measurement that is not actually made.) In order to quantitatively describe a fuzzy observable, we must assume that a measurement is made, and if we do not want our description to change the observable described, we must assume that no measurement is made. We must assign probabilities to the possible outcomes of *unperformed* measurements.

8.3   Upgrading from classical to quantum

The classical probability algorithm cannot accommodate the nontrivial probabilities that are needed for the purpose of defining and quantifying a fuzzy physical property. Nor can the probability algorithms of classical *statistical* physics be used for this purpose, for the nontrivial probabilities of classical physics are distributions over a phase space, and this makes them consistent with the belief that the real (albeit unknown) state is represented by a point $\mathcal{P}$ in this space. In other words, the probabilities of classical statistical physics are (or allow themselves to be thought of as) ignorance probabilities. They enter the picture when relevant facts are ignored, and they disappear, or degenerate into trivial probabilities, when all relevant facts are taken into account. The "uncertainty" relation (4.18), on the other hand, guarantees that quantum-mechanical probabilities cannot be made to disappear.

Arguably the most straightforward way to make room for nontrivial probabilities is to upgrade[1] from a 0-dimensional point $\mathcal{P}$ to a 1-dimensional line $\mathcal{L}$. Instead of representing a probability algorithm by a point in a phase space, we represent it by a 1-dimensional subspace of a vector space $\mathcal{V}$. (Such a subspace is sometimes called a "ray.") And instead of representing elementary tests by subsets of a phase space $\mathcal{S}$, we represent them by the subspaces of $\mathcal{V}$. A 1-dimensional subspace $\mathcal{L}$ can be contained in a subspace $\mathcal{U}$, it can be orthogonal to $\mathcal{U}$ ($\mathcal{L} \perp \mathcal{U}$), but it may be neither contained in nor orthogonal to $\mathcal{U}$; there is a third possibility, and this is what makes room for nontrivial probabilities. $\mathcal{L}$ assigns probability 1 to elementary tests represented by subspaces containing $\mathcal{L}$, it assigns probability 0 to elementary tests represented by subspaces orthogonal to $\mathcal{L}$, and it assigns probabilities greater than 0 and less than 1 to tests represented by subspaces that neither contain nor are orthogonal to $\mathcal{L}$.

8.4   Vector spaces

Using the notation introduced by Paul Dirac, we define a vector space $\mathcal{V}$ as a set of vectors $|a\rangle$, $|b\rangle$, $|c\rangle$... that can be (i) added and (ii) multiplied with real or complex numbers α, β, γ.... The result of either operation is another vector in $\mathcal{V}$. There is a *null vector* $|0\rangle$ such that for any vector $|a\rangle$ we have $|a\rangle + |0\rangle = |a\rangle$ and $0\,|a\rangle = |0\rangle$. If in addition (as we shall assume)

---

[1] The virtual inevitability of this "upgrade" was demonstrated by Jauch (1968).

$\mathcal{V}$ comes equipped with an *inner* or *scalar product*—a machine that accepts two vectors $|a\rangle, |b\rangle$ and returns a real or complex number $\langle a|b\rangle$, then $\mathcal{V}$ is an *inner product space*. In a *real vector space*, only real numbers are admitted, and the scalar product is symmetric as well as linear in both vectors. This means that

$$\langle a|b\rangle = \langle b|a\rangle\,, \tag{8.2}$$

$$\text{if } |A\rangle = \alpha|a\rangle \text{ and } |B\rangle = \beta|b\rangle \text{ then } \langle A|B\rangle = \alpha\beta\langle a|b\rangle\,. \tag{8.3}$$

In a *complex vector space*, these requirements are replaced by

$$\langle a|b\rangle = \langle b|a\rangle^*\,, \tag{8.4}$$

$$\text{if } |A\rangle = \alpha|a\rangle \text{ and } |B\rangle = \beta|b\rangle \text{ then } \langle A|B\rangle = \alpha^*\beta\langle a|b\rangle\,. \tag{8.5}$$

Problem 8.2. *The last two requirements follow from the demand that $\langle a|a\rangle$ be a real number.*

In addition we have that

$$\text{if } |c\rangle = |a\rangle + |b\rangle \text{ then } \langle d|c\rangle = \langle d|a\rangle + \langle d|b\rangle\,, \tag{8.6}$$

$$\langle a|a\rangle \geq 0\,, \tag{8.7}$$

$$\text{if } \langle a|a\rangle = 0 \text{ then } |a\rangle = |0\rangle\,. \tag{8.8}$$

A few more definitions are needed:

- Two vectors are (mutually) *orthogonal* if their scalar product vanishes.
- A set of vectors $|a_1\rangle, \ldots, |a_m\rangle$ is *linearly independent* if $\alpha_1|a_1\rangle + \cdots + \alpha_m|a_m\rangle = 0$ implies that $\alpha_1 = \cdots = \alpha_m = 0$.
- The *dimension* of $\mathcal{V}$ is the maximal number of linearly independent vectors in $\mathcal{V}$. It may be infinite, but only denumerably so.
- A system $|c_1\rangle, |c_2\rangle, \ldots$ of linearly independent vectors is *complete* if every vector $|b\rangle$ in $\mathcal{V}$ can be written as a *linear combination* of these vectors: there are real or complex numbers β_i such that $|b\rangle = \sum_i \beta_i|c_i\rangle$.
- The *norm* of a vector $|a\rangle$ is the positive square root $\sqrt{\langle a|a\rangle}$.
- A *unit vector* is a vector whose norm equals unity.
- A *basis*—short for "orthonormal basis"—is a complete set of mutually orthogonal unit vectors.

Thus if the vectors $|a_i\rangle$ form a basis, then

$$\langle a_i|a_k\rangle = \delta_{ik} \stackrel{\text{Def}}{=} \begin{cases} 0 & \text{if } i \neq k \\ 1 & \text{if } i = k \end{cases}\,, \tag{8.9}$$

and $b_i = \langle a_i|b\rangle$ is the i-th component of $|b\rangle$ with respect to this basis.

Finally, if $\mathcal{V}$ is complex and in addition (as we shall assume) *complete*, it is a *Hilbert space*. Completeness means that every (strongly) convergent sequence $|v_k\rangle$ of elements in $\mathcal{V}$ must have a unique limit $|v\rangle$ in $\mathcal{V}$ [see, e.g., Peres (1995), p. 81].

8.4.1 *Why complex numbers?*

Does the quantum-mechanical probability calculus actually need complex vector spaces, or will real ones do? We obtained the answer in Sec. 5.7, where it was shown that the stability of a free particle requires the amplitude (5.16) associated with a spacetime path to be a phase factor, and thus a complex number. With the present chapter, however, we want to make a fresh start and derive the rules formulated in Sec. 5.1, which went into the derivation of (5.16). To avoid circular reasoning, we could *assume* for now that our vector spaces are complex and decide *after* those rules are in place whether real vector spaces would have been up to the task. But we already know the answer.

8.4.2 *Subspaces and projectors*

Here is what the decomposition of a 3-vector with respect to the basis $(\hat{\mathbf{x}}, \hat{\mathbf{y}}, \hat{\mathbf{z}})$ looks like in the standard notation of Sec. 3.1:

$$\mathbf{v} = \hat{\mathbf{x}}\,(\hat{\mathbf{x}} \cdot \mathbf{v}) + \hat{\mathbf{y}}\,(\hat{\mathbf{y}} \cdot \mathbf{v}) + \hat{\mathbf{z}}\,(\hat{\mathbf{z}} \cdot \mathbf{v}). \tag{8.10}$$

Problem 8.3. *Pencil the vectors* $\hat{\mathbf{x}}\,(\hat{\mathbf{x}} \cdot \mathbf{v})$, $\hat{\mathbf{y}}\,(\hat{\mathbf{y}} \cdot \mathbf{v})$, *and* $\hat{\mathbf{z}}\,(\hat{\mathbf{z}} \cdot \mathbf{v})$ *into Fig. 3.1.*

Here is what the decomposition of a vector with respect to the basis vectors $|a_i\rangle$ looks like in Dirac's notation:

$$|v\rangle = \sum_i |a_i\rangle\langle a_i|v\rangle. \tag{8.11}$$

The advantage of this notation is that it consolidates two ways of thinking about the terms on the right-hand side, inasmuch as they may be split in either of two ways:

$$\text{(i)}\quad |a_i\rangle\langle a_i|\,|v\rangle, \quad \text{(ii)}\quad |a_i\rangle\,\langle a_i|v\rangle. \tag{8.12}$$

$|v\rangle$ and $|a_i\rangle$ are vectors. $\langle a_i|v\rangle$ is the i-the component of $|v\rangle$ with respect to the basis vectors $|a_i\rangle$. And $|a_i\rangle\langle a_i|$ is a machine that accepts a vector and returns a vector: insert $|v\rangle$, and out pops $|a_i\rangle\langle a_i|v\rangle$—the vector $|a_i\rangle$ multiplied by $\langle a_i|v\rangle$. We refer to it as the *projection* of $|v\rangle$ into the 1-dimensional subspace containing $|a_i\rangle$, and we refer to an operator that projects into a subspace as a *projector*. Think of (i) as what we do—projecting $|v\rangle$ into the 1-dimensional subspace containing $|a_i\rangle$—and of (ii) as what we get as a result—the projection of $|v\rangle$ into this subspace.

A *subspace* is a vector space within a vector space. Specifically, $\mathcal{U}$ is a subspace of $\mathcal{V}$ if and only if it satisfies the following condition: if $|u\rangle$ and $|v\rangle$ are in $\mathcal{U}$ and a and b are numbers, then $a|u\rangle + b|v\rangle$ is in $\mathcal{U}$. If a subspace $\mathcal{U}$ has a dimension that is (denumerably) infinite, we shall tacitly assume that it is closed: every sequence of vectors in $\mathcal{U}$ that converges in $\mathcal{V}$—i.e., that has as its limit a vector in $\mathcal{V}$—also converges in $\mathcal{U}$.

There is a one-to-one correspondence between subspaces and projectors. If $\mathcal{U}$ is a subspace, so that there is in $\mathcal{V}$ a set of mutually orthogonal unit vectors $|b_1\rangle, \ldots, |b_m\rangle$ that allows every vector in $\mathcal{U}$ to be written as a linear combination $\sum_{k=1}^{m} v_k |b_k\rangle$, then there is a unique operator that projects into $\mathcal{U}$, namely,

$$\hat{\mathbf{P}}(\mathcal{U}) = \sum_{k=1}^{m} |b_k\rangle\langle b_k|. \tag{8.13}$$

We say that $\mathcal{U}$ is *spanned* by the vectors $|b_1\rangle, \ldots, |b_m\rangle$. While the projector (8.13) is unique, the vectors that span $\mathcal{U}$ are not. If $|c_1\rangle, \ldots, |c_m\rangle$ is a different set of mutually orthogonal unit vectors that span $\mathcal{U}$, then

$$\sum_{k=1}^{m} |c_k\rangle\langle c_k| = \sum_{k=1}^{m} |b_k\rangle\langle b_k|. \tag{8.14}$$

Problem 8.4. *If $|v\rangle \subset \mathcal{U}$ then $\hat{\mathbf{P}}(\mathcal{U})\,|v\rangle = |v\rangle$.*

Problem 8.5. $\hat{\mathbf{P}}(\mathcal{U})\,\hat{\mathbf{P}}(\mathcal{U}) = \hat{\mathbf{P}}(\mathcal{U})$.

If the vectors $|a_k\rangle$ $(k = 1, \ldots, n)$ constitute a basis in $\mathcal{V}$, then

$$\hat{\mathbf{I}} = \sum_{k=1}^{n} |a_k\rangle\langle a_k| \tag{8.15}$$

is the *identity operator*, which projects every vector in $\mathcal{V}$ into $\mathcal{V}$—a long-winded way of saying that it does nothing.

Problem 8.6. *The projector $|v\rangle\langle v|$ is invariant under the phase transformation $|v\rangle \rightarrow e^{i\alpha}|v\rangle$.*

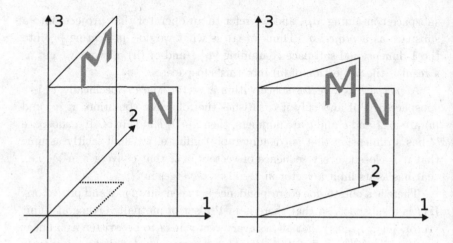

Fig. 8.2   Two 2-dimensional subspaces of a 3-dimensional (real) vector space. Only in the left diagram do the corresponding projectors commute.

8.4.3   *Commuting and non-commuting projectors*

Two operators $\hat{A}, \hat{B}$ are said to *commute* if

$$\hat{A}\hat{B} = \hat{B}\hat{A}. \tag{8.16}$$

This is the shorter way of saying that $\hat{A}\hat{B}|v\rangle = \hat{B}\hat{A}|v\rangle$ for every vector $|v\rangle$. It is not hard to see that two projectors $\hat{M}$ and $\hat{N}$ commute if and only if there is a basis $|a_i\rangle$ such that

$$\hat{M} = \sum_{\text{some } i} |a_i\rangle\langle a_i| \quad \text{and} \quad \hat{N} = \sum_{\text{some } k} |a_k\rangle\langle a_k|. \tag{8.17}$$

Consider the following example (Fig. 8.2):

$$\hat{M} = |2\rangle\langle 2| + |3\rangle\langle 3|, \qquad \hat{N} = |1\rangle\langle 1| + |3\rangle\langle 3|. \tag{8.18}$$

Problem 8.7. (∗) *The two projectors commute if there is a basis to which the vectors $|1\rangle, |2\rangle, |3\rangle$ belong.*

If $|1\rangle$ and $|2\rangle$ are not orthogonal (while the remaining pairs are), we have

$$\hat{M}\hat{N} = |2\rangle\langle 2|1\rangle\langle 1| + |3\rangle\langle 3|, \qquad \hat{N}\hat{M} = |1\rangle\langle 1|2\rangle\langle 2| + |3\rangle\langle 3|, \tag{8.19}$$

that is, $\hat{M}$ and $\hat{N}$ do not commute. Figure 8.3 illustrates the respective actions of $\hat{M}\hat{N}$ and $\hat{N}\hat{M}$ on a vector $|v\rangle$.

Our next task is to translate the commutativity condition for projectors into the language of subspaces. The following two definitions will help us do this.

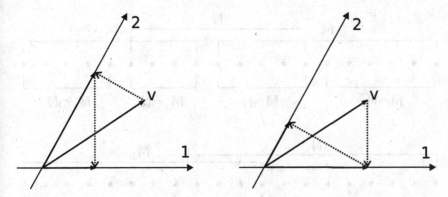

Fig. 8.3   The situation illustrated in Fig. 8.2 (right diagram) seen from above, looking down the 3-axis.   On the left, $|v\rangle$ is projected into $\mathcal{M}$, then the resulting vector is projected into $\mathcal{N}$.   On the right, $|v\rangle$ is projected into $\mathcal{N}$, then the resulting vector is projected into $\mathcal{M}$.

- The *orthocomplement* $\mathcal{A}_\perp$ of a subspace $\mathcal{A}$ contains all vectors that are orthogonal to all vectors in $\mathcal{A}$.
- The *span* $\mathcal{A} \cup \mathcal{B}$ of two subspaces $\mathcal{A}$ and $\mathcal{B}$ is the smallest subspace that contains both $\mathcal{A}$ and $\mathcal{B}$.

Problem 8.8. $\mathcal{A}_\perp$ *is a subspace.*

Problem 8.9. *The intersection* $\mathcal{A} \cap \mathcal{B}$ *of two subspaces, which contains all vectors that are contained in both subspaces, is a subspace.*

Problem 8.10. $(\mathcal{A}_\perp)_\perp = \mathcal{A}.$

Problem 8.11. *Convince yourself that* $\mathcal{A} \subset \mathcal{B}_\perp$ *and* $\mathcal{B} \subset \mathcal{A}_\perp$ *are equivalent.*

Figure 8.4 illustrates the relations that hold between $\mathcal{M}$, $\mathcal{N}$, their orthocomplements, and the intersections of all these just in case the corresponding projectors $\hat{\mathbf{M}}, \hat{\mathbf{N}}$ commute. We gather from it that

$$(\mathcal{M} \cap \mathcal{N}) \cup (\mathcal{M} \cap \mathcal{N}_\perp) \cup (\mathcal{M}_\perp \cap \mathcal{N}) \cup (\mathcal{M}_\perp \cap \mathcal{N}_\perp) = \mathcal{V}. \qquad (8.20)$$

This is the wanted expression of the commutativity condition for projectors in terms of the corresponding subspaces. A quick check will confirm that it is satisfied by the subspaces in the left half of Fig. 8.2: while $\mathcal{M}_\perp \cap \mathcal{N}_\perp$ is empty, $\mathcal{M}_\perp \cap \mathcal{N}$ is the subspace containing $|1\rangle$, $\mathcal{M} \cap \mathcal{N}_\perp$ is the subspace containing $|2\rangle$, and $\mathcal{M} \cap \mathcal{N}$ is the subspace containing $|3\rangle$. On the other hand, Eq. (8.20) does not hold for the subspaces in the right half of Fig. 8.2:

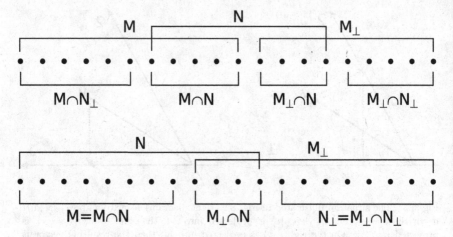

Fig. 8.4　The relations that hold between $\mathcal{M}$, $\mathcal{N}$, their orthocomplements, and the intersections of all these if the corresponding projectors can be written in terms of a common basis, as in Eq. (8.17). The dots represent the basis vectors. The horizontal bracket associated with a subspace groups the 1-dimensional projectors that make up the projector into that subspace. Above: neither of the subspaces $\mathcal{M}$, $\mathcal{N}$ is contained in the other. Below: $\mathcal{M} \subset \mathcal{N}$.

while $\mathcal{M} \cap \mathcal{N}$ is the 1-dimensional subspace containing $|3\rangle$, the intersections $\mathcal{M} \cap \mathcal{N}_\perp$, $\mathcal{M}_\perp \cap \mathcal{N}$, and $\mathcal{M}_\perp \cap \mathcal{N}_\perp$ are all empty.

8.5　Compatible and incompatible elementary tests

This brings us back to our central theme, the stability of matter. The stability of atoms requires an inequality of the form (4.18); the product of the "uncertainties" associated with a relative position and the corresponding relative momentum must have a positive lower limit. This makes it impossible to simultaneously measure both quantities with arbitrary precision. In other words, the stability of matter requires, via the stability of atoms, that the two quantities be *incompatible*.

To arrive at a formal definition of compatibility, we consider two elementary tests. We first assume that one of the possible outcomes of the second test (say, the positive outcome $\mathcal{N}$) is implied by one of the possible outcomes of the first test (say, the positive outcome $\mathcal{M}$): whenever the outcome $\mathcal{M}$ is obtained, a subsequent test of $\mathcal{N}$ is bound to yield a positive result.

Problem 8.12. *This is equivalent to $\mathcal{M} \subset \mathcal{N}$.*

Problem 8.13. *If $\mathcal{N}$ represents the positive outcome of an elementary test, $\mathcal{N}_\perp$ represents the negative outcome.*

There are three possible combinations of outcomes: (i) $\mathcal{M}$ and $\mathcal{N}$, (ii) $\mathcal{M}_\perp$ and $\mathcal{N}$, (iii) $\mathcal{M}_\perp$ and $\mathcal{N}_\perp$. Requiring compatibility for the two tests amounts to requiring that for each combination of outcomes there should be an algorithm assigning to it a probability equal to 1. In other words, the two tests are compatible if and only if there are three 1-dimensional subspaces $\mathcal{L}_i$ such that

 (i) $\mathcal{L}_1 \subset \mathcal{M} \cap \mathcal{N}$: $\mathcal{L}_1$ assigns probability 1 to both $\mathcal{M}$ and $\mathcal{N}$,
 (ii) $\mathcal{L}_2 \subset \mathcal{M}_\perp \cap \mathcal{N}$: $\mathcal{L}_2$ assigns probability 1 to both $\mathcal{M}_\perp$ and $\mathcal{N}$,
(iii) $\mathcal{L}_3 \subset \mathcal{M}_\perp \cap \mathcal{N}_\perp$: $\mathcal{L}_3$ assigns probability 1 to both $\mathcal{M}_\perp$ and $\mathcal{N}_\perp$.

It is easy to see that these requirements are satisfied: $\mathcal{M} \subset \mathcal{N}$, therefore $\mathcal{L}_1$ exists; $\mathcal{N} \neq \mathcal{M}$, so $\mathcal{N}$ contains a line orthogonal to $\mathcal{M}$, therefore $\mathcal{L}_2$ exists; and because $\mathcal{M} \subset \mathcal{N}$ implies $\mathcal{N}_\perp \subset \mathcal{M}_\perp$ (Problem 8.11), therefore $\mathcal{L}_3$ exists. Thus if either outcome of the first test implies either outcome of the second test, the two tests are compatible.

If neither outcome of the first test implies either outcome of the second test, then compatibility requires, in addition, the existence of a 1-dimensional subspace $\mathcal{L}_4$ such that

(iv) $\mathcal{L}_4 \subset \mathcal{M} \cap \mathcal{N}_\perp$: $\mathcal{L}_4$ assigns probability 1 to both $\mathcal{M}$ and $\mathcal{N}_\perp$.

If these four requirements are satisfied, none of the intersections $\mathcal{M} \cap \mathcal{N}$, $\mathcal{M} \cap \mathcal{N}_\perp$, $\mathcal{M}_\perp \cap \mathcal{N}$, and $\mathcal{M}_\perp \cap \mathcal{N}_\perp$ can be equal to $\emptyset$. Since these intersections are mutually orthogonal, one can find a set of mutually orthogonal unit vectors $|a_i\rangle$ $(i = 1, \ldots, n)$ such that those with $i = 1, \ldots, j$ span $\mathcal{M} \cap \mathcal{N}$, those with $i = j + 1, \ldots, k$ span $\mathcal{M} \cap \mathcal{N}_\perp$, those with $i = k + 1, \ldots, m$ span $\mathcal{M}_\perp \cap \mathcal{N}$, and those with $i = m + 1, \ldots, n$ span $\mathcal{M}_\perp \cap \mathcal{N}_\perp$. To establish that the n vectors $|a_i\rangle$ in fact constitute a basis, we assume the contrary. We assume, in other words, that there is a 1-dimensional subspace $\mathcal{L}$ that is orthogonal to each of those four intersections. But this is the same as saying that there is a probability algorithm that assigns probability 0 to all possible combinations of outcomes, and this is a *reductio ad absurdum* of our assumption.

The bottom line: two elementary tests with respective positive outcomes $\mathcal{M}$ and $\mathcal{N}$ are compatible if and only if Eq. (8.20) holds. (If $\mathcal{M} \subset \mathcal{N}$, then this equation holds with $\mathcal{M} \cap \mathcal{N}_\perp = \emptyset$.)

8.6   Noncontextuality

If the subspaces $\mathcal{A}$ and $\mathcal{B}$ represent measurement outcomes, what measurement outcome is represented by the span $\mathcal{A} \cup \mathcal{B}$?

We begin with the following observations. Let $p(\mathcal{A})$ and $p(\mathcal{B})$ be the respective probabilities of obtaining the outcomes represented by $\mathcal{A}$ and $\mathcal{B}$. Because a 1-dimensional subspace that is contained in either $\mathcal{A}$ or $\mathcal{B}$ is contained in $\mathcal{A} \cup \mathcal{B}$, we have that

$$[p(\mathcal{A}) = 1 \text{ or } p(\mathcal{B}) = 1] \implies p(\mathcal{A} \cup \mathcal{B}) = 1. \tag{8.21}$$

In words: if either $\mathcal{A}$ or $\mathcal{B}$ has probability 1, then $\mathcal{A} \cup \mathcal{B}$ has probability 1. Again, because a line orthogonal to both $\mathcal{A}$ and $\mathcal{B}$ is orthogonal to $\mathcal{A} \cup \mathcal{B}$, we also have that

$$[p(\mathcal{A}) = 0 \text{ and } p(\mathcal{B}) = 0] \implies p(\mathcal{A} \cup \mathcal{B}) = 0. \tag{8.22}$$

The implications (8.21) and (8.22) hold, in particular, if $\mathcal{A}$ and $\mathcal{B}$ represent disjoint intervals A and B of a variable Q with a continuous range of values. But note that a 1-dimensional subspace can be in $\mathcal{A} \cup \mathcal{B}$ without being contained in either $\mathcal{A}$ or $\mathcal{B}$. This means that the outcome $\mathcal{A} \cup \mathcal{B}$ can be certain even if neither $\mathcal{A}$ nor $\mathcal{B}$ is certain. Obtaining the outcome $\mathcal{A} \cup \mathcal{B}$ therefore does not imply that the value of Q is either A or B. *A fortiori*, it does not imply that the value of Q is any smaller interval in either A or B, let alone a definite number.

Next imagine two perfect—one hundred percent efficient—detectors $D(A)$ and $D(B)$ monitoring the two intervals. If the probabilities $p(\mathcal{A})$ and $p(\mathcal{B})$ are both greater than 0 (and therefore less than 1), then it is not certain that $D(A)$ will click, and it is not certain that $D(B)$ will click. Yet if $p(\mathcal{A} \cup \mathcal{B}) = 1$, then it *is* certain that either $D(A)$ or $D(B)$ will click. What makes this certain?

The answer lies in the fact that quantum-mechanical probability assignments are invariably made on the (tacit) assumption that a measurement is successfully made; there is an outcome. For instance, if A and B are disjoint regions of space, and if a measurement has indicated the presence of a particle in the union $A \cup B$ of these regions, then the tacit assumption is that a subsequent position measurement made with two detectors monitoring the respective regions A and B will yield an outcome—either detector will click. So there is no mystery here, but the implication is that quantum mechanics only gives us probabilities with which this or that outcome is obtained in a *successful* measurement. It does not give us the probability with which an attempted measurement will succeed. *A fortiori*, it is incapable of formulating sufficient conditions for the success of a measurement [Ulfbeck and Bohr (2001)].

Now consider the following two measurements. The first, M_1, has three possible outcomes: $\mathcal{A}$, $\mathcal{B}$, and $\mathcal{C}$. The second, M_2, has two: $\mathcal{A} \cup \mathcal{B}$ and $\mathcal{C}$. We therefore have

$$p(\mathcal{A}) + p(\mathcal{B}) + p(\mathcal{C}) = 1 \quad \text{as well as} \quad p(\mathcal{A} \cup \mathcal{B}) + p(\mathcal{C}) = 1. \qquad (8.23)$$

If, for instance, A, B, and C are disjoint regions of space, and if a particle is certain to be found in the union of these regions, then M_1 is certain to find it in one of these three regions, while M_2 is certain to find it in either $A \cup B$ or C. It follows that

$$p(\mathcal{A} \cup \mathcal{B}) = p(\mathcal{A}) + p(\mathcal{B}). \qquad (8.24)$$

Or does it? Both measurements test for the possession of C, but whereas M_1 includes two additional tests, M_2 only includes one. What gives us the right to assume that the probability of a possible measurement outcome is independent of what the other possible outcomes are? Equations (8.23) hold in different experimental contexts. Is it legitimate to separate them from their specific contexts so as to draw the conclusion (8.24)? In other words, is it legitimate to assume *noncontextuality*?

Let us remember our objective. We are looking for a probability algorithm that is capable of accommodating nontrivial probabilities and incompatible elementary tests. If common sense in the form of noncontextuality is consistent with these requirements, we go for it. No need to make the world stranger than it already is. By hindsight we know that Nature concurs.[2]

Problem 8.14. *Projectors are said to be (mutually) orthogonal if the corresponding subspaces are. Show that the sum of two orthogonal projectors is a projector.*

Problem 8.15. *If the orthogonal projectors $\hat{\mathbf{A}}$ and $\hat{\mathbf{B}}$ project into $\mathcal{A}$ and $\mathcal{B}$, respectively, the sum $\hat{\mathbf{A}} + \hat{\mathbf{B}}$ projects into $\mathcal{A} \cup \mathcal{B}$.*

Problem 8.16. *The projectors $\hat{\mathbf{A}}$ and $\hat{\mathbf{B}}$ corresponding to different outcomes $\mathcal{A}$ and $\mathcal{B}$ of the same measurement are orthogonal.*

---

[2]We also know that contextuality is an inescapable feature of situations in which probabilities are assigned either on the basis of past *and* future outcomes (Sec. 13.8.1) or to outcomes of measurements performed on entangled systems (Sec. 13.4).

8.7　The core postulates

We are now ready to assemble the core postulates of quantum mechanics. The first was stated in Sec. 8.3:

Postulate 1. *Measurement outcomes are represented by the projectors of a vector space.*

In Sec. 8.5 we concluded that two elementary tests $\mathcal{M}$ and $\mathcal{N}$ are compatible if and only if Eq. (8.20) holds, and in Sec. 8.4.3 we found that this is the case if and only if the corresponding projectors commute. Hence

Postulate 2. *The outcomes of compatible elementary tests correspond to commuting projectors.*

From the previous section, finally, we derive the following:

Postulate 3. *If $\hat{\mathbf{A}}$ and $\hat{\mathbf{B}}$ are orthogonal projectors, then the probability of the outcome represented by $\hat{\mathbf{A}} + \hat{\mathbf{B}}$ is the sum of the probabilities of the outcomes represented by $\hat{\mathbf{A}}$ and $\hat{\mathbf{B}}$, respectively.*

8.8　The trace rule

Postulates 1–3 are sufficient [Peres, 1995, p. 190] to prove Gleason's theorem [Gleason (1957); Pitowsky (1998); Cooke *et al.* (1985)], which holds for vector spaces with at least three dimensions.[3] The theorem states that the probability of obtaining the outcome represented by the projector $\hat{\mathbf{P}}$ is given by

$$p(\hat{\mathbf{P}}) = \mathrm{Tr}(\hat{\mathbf{W}}\hat{\mathbf{P}}),　\qquad (8.25)$$

where $\hat{\mathbf{W}}$ is a unique operator, known as *density operator*, whose properties will be listed in a moment. The *trace* of an operator $\hat{\mathbf{X}}$ is defined by

$$\mathrm{Tr}(\hat{\mathbf{X}}) = \sum_i \langle a_i | \hat{\mathbf{X}} | a_i \rangle,　\qquad (8.26)$$

---

[3]More recently, the validity of Gleason's theorem has been extended to include 2-dimensional vector spaces [Fuchs (2001); Busch (2003); Caves *et al.* (2004)] by generalizing from the projector valued measures used in this book to positive operator valued measures or POVMs [Peres, 1995, Sec. 9–5]. If outcomes are represented by positive operators, assigning probability 1 to one outcome no longer necessitates assigning probability 0 to a different outcome of the same measurement. POVMs are thus suited for dealing with measurements that have overlapping outcomes, such as position measurements with detectors whose sensitive regions intersect.

where the vectors $|a_i\rangle$ form a basis and the summation extends from $i = 1$ to the dimension of the system's vector space. Think of $\langle a_i|\hat{\mathbf{X}}|a_i\rangle$ as the i-th component of the vector $\hat{\mathbf{X}}|a_i\rangle$ with respect to this basis. The definition of the trace appears to be contingent on a basis, but actually this is not the case. Let us insert a couple of identity operators $\hat{\mathbf{I}} = \sum_i |b_i\rangle\langle b_i|$:

$$\sum_i \langle a_i|\hat{\mathbf{X}}|a_i\rangle = \sum_i \langle a_i|\hat{\mathbf{I}}\hat{\mathbf{X}}\hat{\mathbf{I}}|a_i\rangle = \sum_i \sum_j \sum_k \langle a_i|b_j\rangle\langle b_j|\hat{\mathbf{X}}|b_k\rangle\langle b_k|a_i\rangle .$$

Now we rearrange the complex factors on the right-hand side and take out two identity operators:

$$\sum_i \sum_j \sum_k \langle b_k|a_i\rangle\langle a_i|b_j\rangle\langle b_j|\hat{\mathbf{X}}|b_k\rangle = \sum_j \sum_k \langle b_k|b_j\rangle\langle b_j|\hat{\mathbf{X}}|b_k\rangle = \sum_k \langle b_k|\hat{\mathbf{X}}|b_k\rangle .$$

Voilà: the trace is independent of the basis that has been used in its definition.

Here is the promised list of properties of the density operator:

(i) $\hat{\mathbf{W}}$ is *linear*: $\hat{\mathbf{W}}\big(\alpha|a\rangle + \beta|b\rangle\big) = \alpha\,\hat{\mathbf{W}}|a\rangle + \beta\,\hat{\mathbf{W}}|b\rangle$.
(ii) $\hat{\mathbf{W}}$ is *self-adjoint*: $\langle a|\hat{\mathbf{W}}|b\rangle = \langle b|\hat{\mathbf{W}}|a\rangle^*$.
(iii) $\hat{\mathbf{W}}$ is *positive*: $\langle a|\hat{\mathbf{W}}|a\rangle \geq 0$.
(iv) $\mathrm{Tr}(\hat{\mathbf{W}}) = 1$.
(v) $\hat{\mathbf{W}}^2 \leq \hat{\mathbf{W}}$.

Let us try to understand why $\hat{\mathbf{W}}$ has these properties.

Problem 8.17. *If $\hat{\mathbf{P}}$ is a 1-dimensional projector $|v\rangle\langle v|$, the trace rule (8.25) reduces to $p(\hat{\mathbf{P}}) = \langle v|\hat{\mathbf{W}}|v\rangle$.*

The fact that $\hat{\mathbf{W}}$ is self-adjoint ensures that the probability $\langle v|\hat{\mathbf{W}}|v\rangle$ is a real number. (What could be the meaning of a complex probability?)

The positivity of $\hat{\mathbf{W}}$ ensures that $\langle v|\hat{\mathbf{W}}|v\rangle$ does not come out negative. (What could be the meaning of a negative probability?)

The outcomes of a *maximal test* (also known as a *complete measurement*) are represented by a complete system of (mutually orthogonal) 1-dimensional projectors $|a_k\rangle\langle a_k|$ ($k = 1,\ldots,n$). $\mathrm{Tr}(\hat{\mathbf{W}}) = 1$ ensures that the corresponding probabilities $\langle a_i|\hat{\mathbf{W}}|a_i\rangle$ add up to 1. Together with the positivity of $\hat{\mathbf{W}}$, this ensures that no probability comes out greater than 1. (What could be the meaning of a probability greater than 1?)

The last property covers two possibilities, $\hat{\mathbf{W}}^2 = \hat{\mathbf{W}}$ and $\hat{\mathbf{W}}^2 < \hat{\mathbf{W}}$. These will be discussed after a short mathematical digression.

8.9   Self-adjoint operators and the spectral theorem

In what follows we take the liberty of writing $|\hat{\mathbf{A}}b\rangle$ instead of $\hat{\mathbf{A}}|b\rangle$ when referring to the vector returned by $\hat{\mathbf{A}}$ upon insertion of $|b\rangle$.

If $\hat{\mathbf{A}}$ is a linear operator (like $\hat{\mathbf{W}}$ above), there exists a linear operator $\hat{\mathbf{A}}^\dagger$, called the *adjoint* of $\hat{\mathbf{A}}$, such that $\langle a|\hat{\mathbf{A}}b\rangle = \langle \hat{\mathbf{A}}^\dagger a|b\rangle$ for every pair of vectors $|a\rangle, |b\rangle$.

Problem 8.18. $\langle a|\hat{\mathbf{A}}b\rangle = \langle b|\hat{\mathbf{A}}^\dagger a\rangle^*$.

Problem 8.19. *Show that (i)* $(\hat{\mathbf{A}}^\dagger)^\dagger = \hat{\mathbf{A}}$ *and (ii)* $(\hat{\mathbf{A}}\hat{\mathbf{B}})^\dagger = \hat{\mathbf{B}}^\dagger\hat{\mathbf{A}}^\dagger$.

If (i) $\hat{\mathbf{A}}$ is linear, (ii) $|a\rangle$ is a vector, and (iii) α is a complex number such that $\hat{\mathbf{A}}|a\rangle = \alpha|a\rangle$, then $|a\rangle$ is an *eigenvector* of $\hat{\mathbf{A}}$ and α is the corresponding *eigenvalue*.

Problem 8.20. *A projector has exactly two eigenvalues, namely....*

Saying that $\hat{\mathbf{A}}$ is *self-adjoint* is the same as saying that $\hat{\mathbf{A}}^\dagger = \hat{\mathbf{A}}$. Self-adjoint operators have two important properties. For one, their eigenvalues are real, as we gather from the following equations:

$$\alpha\langle a|a\rangle = \langle a|\hat{\mathbf{A}}a\rangle = \langle \hat{\mathbf{A}}a|a\rangle = \alpha^*\langle a|a\rangle. \tag{8.27}$$

For another, the eigenvectors of a self-adjoint operator that correspond to different eigenvalues $\alpha_1 \neq \alpha_2$ are orthogonal, as we gather from these equations:

$$0 = \langle a_1|\hat{\mathbf{A}}a_2\rangle - \langle \hat{\mathbf{A}}a_1|a_2\rangle = (\alpha_2 - \alpha_1)\langle a_1|a_2\rangle. \tag{8.28}$$

Problem 8.21. *If a linear operator $\hat{\mathbf{A}}$ has $m > 1$ linearly independent eigenvectors with the same eigenvalue, then the eigenvectors of $\hat{\mathbf{A}}$ corresponding to this eigenvalue form an m-dimensional subspace.*

For every self-adjoint operator $\hat{\mathbf{A}}$, it is possible to construct a basis made up entirely of eigenvectors of $\hat{\mathbf{A}}$ [see, for example, Marchildon, 2002, Sec. 2.5]. Suppose that the vectors $|a_i\rangle$ constitute such a basis, and that the corresponding eigenvalues are α_i. Define the operator $\hat{\mathbf{A}}' = \sum_i \alpha_i |a_i\rangle\langle a_i|$ and insert into it the generic vector $|v\rangle = \sum_j v_j|a_j\rangle$:

$$\hat{\mathbf{A}}'|v\rangle = \sum_i \sum_j \alpha_i v_j |a_i\rangle\langle a_i|a_j\rangle = \sum_i \alpha_i v_i|a_i\rangle. \tag{8.29}$$

We would get the same result if we inserted $|v\rangle$ into $\hat{\mathbf{A}}$. But if two operators act in the same way on every vector, they are identical: $\hat{\mathbf{A}}' = \hat{\mathbf{A}}$. Hence the *spectral theorem*, which says that every self-adjoint operator $\hat{\mathbf{A}}$ has a *spectral decomposition*

$$\hat{\mathbf{A}} = \sum_i \alpha_i |a_i\rangle\langle a_i|. \tag{8.30}$$

The eigenvalues α_i form the *spectrum* of $\hat{\mathbf{A}}$.

Problem 8.22. *Write down the spectral decomposition of a projector $\hat{\mathbf{P}}$.*

Problem 8.23. (∗) *A self-adjoint operator satisfying $\hat{\mathbf{A}}^2 = \hat{\mathbf{A}}$ is a projector.*

Problem 8.24. *The trace of a projector is the dimension of the subspace into which it projects.*

8.10 Pure states and mixed states

If the density operator is *idempotent* (i.e., $\hat{\mathbf{W}}^2 = \hat{\mathbf{W}}$), it is a projector, and since the trace of $\hat{\mathbf{W}}$ equals unity, it projects into a 1-dimensional subspace. Since we began by upgrading from a point in a phase space to a line (or ray) in a vector space, we are not surprised by this result. In this case $\hat{\mathbf{W}} = |w\rangle\langle w|$ is called a *pure state*, $|w\rangle$ is referred to as a *state vector* (which is unique up to a phase factor), and the trace rule simplifies to

$$p(\hat{\mathbf{P}}) = \langle w|\hat{\mathbf{P}}|w\rangle. \tag{8.31}$$

If in addition $\hat{\mathbf{P}} = |v\rangle\langle v|$, the trace rule boils down to the *Born rule*,

$$p(\hat{\mathbf{P}}) = \langle w|v\rangle\langle v|w\rangle = |\langle v|w\rangle|^2. \tag{8.32}$$

If $\hat{\mathbf{W}}^2 < \hat{\mathbf{W}}$, which means that $\langle a|\hat{\mathbf{W}}^2|a\rangle < \langle a|\hat{\mathbf{W}}|a\rangle$ for every vector $|a\rangle$, then $\hat{\mathbf{W}}$ is called a *mixed state* or *mixture*. What do we make of this possibility? Given a spectral decomposition of $\hat{\mathbf{W}}$, the inequality $\hat{\mathbf{W}}^2 < \hat{\mathbf{W}}$ takes the form

$$\sum_k \alpha_k^2 |a_k\rangle\langle a_k| < \sum_k \alpha_k |a_k\rangle\langle a_k|. \tag{8.33}$$

Problem 8.25. (∗) *For all eigenvalues α_k of a mixture, $0 < \alpha_k < 1$.*

Problem 8.26. (∗) $\sum_k \alpha_k = 1$.

The eigenvalues of a mixed state thus have all the properties one expects from nontrivial probabilities associated with mutually exclusive and jointly exhaustive possibilities: they are real, they are greater than 0, they are less than 1, and they add up to 1. The obvious conclusion is that mixed states define probability distributions over probability distributions. They add a second layer of uncertainty to that inherent in pure states. There are situations in which this additional uncertainty is subjective in the same sense in which probability distributions over a classical phase space are subjective: the uncertainty arises from a lack of knowledge of relevant facts. But there are also situations in which the additional uncertainty is due to the nonexistence of relevant facts. In these situations it represents an additional objective fuzziness, over and above that associated with the individual projectors $|a_k\rangle\langle a_k|$ in Eq. (8.33). We will encounter some examples of this kind of uncertainty in later chapters.

8.11 How probabilities depend on measurement outcomes

It bears repetition: quantum mechanics serves to assign probabilities to possible measurement outcomes on the basis of actual outcomes. What we have learned so far in this chapter is (i) that the probabilities of possible measurement outcomes are encoded in a density operator and (ii) how they can be extracted from it. Our next order of business is to find out how the density operator is determined by actual outcomes.

Suppose that $\hat{\mathbf{W}}_1$ is the density operator appropriate for assigning probabilities to the possible outcomes of any measurement that may be made at the time t_1. And suppose that a measurement M is made at t_1, and that the outcome obtained is represented by the projector $\hat{\mathbf{P}}$. What is the density operator $\hat{\mathbf{W}}_2$ appropriate for assigning probabilities to the possible outcomes of whichever measurement may be made next, at $t_2 > t_1$?

As is customary in discussions of this kind, we focus on repeatable measurements. If a physical system is subjected to two consecutive identical measurements, and if the second measurement invariably yields the same outcome as the first, we call these measurements *repeatable*.[4]

---

[4]This definition departs from the standard definition in that it allows $\hat{\mathbf{W}}_2$ to be independent of the time interval between the two measurements. (The standard definition assumes that the second measurement is performed "immediately" after the first.)

Let us put together the relevant information at hand:

(i) $\hat{\mathbf{W}}_2$ is constructed out of $\hat{\mathbf{W}}_1$ and $\hat{\mathbf{P}}$.

(ii) $\text{Tr}(\hat{\mathbf{W}}_2\hat{\mathbf{P}}_\vdash) = 0$, where $\hat{\mathbf{P}}_\vdash$ is any possible outcome of M other than $\hat{\mathbf{P}}$.

(iii) $\hat{\mathbf{W}}_2$ is self-adjoint.

(iv) $\text{Tr}(\hat{\mathbf{W}}_2\hat{\mathbf{P}}) = 1$.

Problem 8.27. $\hat{\mathbf{W}}\hat{\mathbf{P}}$ *is self-adjoint if and only if* $\hat{\mathbf{W}}$ *and* $\hat{\mathbf{P}}$ *commute.*

The first two conditions are satisfied by $\hat{\mathbf{W}}_2 = \hat{\mathbf{W}}_1\hat{\mathbf{P}}$, $\hat{\mathbf{W}}_2 = \hat{\mathbf{P}}\hat{\mathbf{W}}_1$, and $\hat{\mathbf{W}}_2 = \hat{\mathbf{P}}\hat{\mathbf{W}}_1\hat{\mathbf{P}}$, but only the last of these candidates satisfies the third condition in case $\hat{\mathbf{W}}_1$ and $\hat{\mathbf{P}}$ fail to commute. To satisfy the final condition as well, all we have to do is divide by $\text{Tr}(\hat{\mathbf{W}}_1\hat{\mathbf{P}}) = \text{Tr}(\hat{\mathbf{P}}\hat{\mathbf{W}}_1) = \text{Tr}(\hat{\mathbf{P}}\hat{\mathbf{W}}_1\hat{\mathbf{P}})$. Thus,

$$\hat{\mathbf{W}}_2 = \frac{\hat{\mathbf{P}}\hat{\mathbf{W}}_1\hat{\mathbf{P}}}{\text{Tr}(\hat{\mathbf{W}}_1\hat{\mathbf{P}})}\,. \tag{8.34}$$

Now suppose that M is a maximal test, and that $\hat{\mathbf{P}} = |w\rangle\langle w|$. Then

$$\hat{\mathbf{W}}_2 = \frac{|w\rangle\langle w|\hat{\mathbf{W}}_1|w\rangle\langle w|}{\text{Tr}(\hat{\mathbf{W}}_1|w\rangle\langle w|)} = |w\rangle\frac{\langle w|\hat{\mathbf{W}}_1|w\rangle}{\langle w|\hat{\mathbf{W}}_1|w\rangle}\langle w| = |w\rangle\langle w|\,. \tag{8.35}$$

Lo and behold, if we update the density operator to take into account the outcome of a maximal test, it turns into the very projector that represents this outcome. Observe that in this case $\hat{\mathbf{W}}_2$ is independent of $\hat{\mathbf{W}}_1$.

8.12 How probabilities depend on the times of measurements

We now relax the condition of repeatability and require only that measurements be verifiable. A measurement M_1, performed at the time t_1, is *verifiable* if it is possible to confirm its outcome by a measurement M_2 performed at the time $t_2 > t_1$. If M_1 is repeatable, then it can be verified by simply repeating it. Otherwise the verification requires (i) the performance of a *different* measurement and (ii) the existence of a one-to-one correspondence between the possible outcomes of M_1 and those of M_2. The second requirement makes it possible to infer the outcome of M_1 from the outcome of M_2.

If the two measurements are maximal tests, then there are two bases, one made up of the vectors $|a_i\rangle$, the other made up of the vectors $|b_k\rangle$, such that the projectors $|a_i\rangle\langle a_i|$ represent the possible outcomes of M_1 and the projectors $|b_k\rangle\langle b_k|$ represent the possible outcomes of M_2. The two bases are related by some transformation

$$|a_k\rangle \longrightarrow |b_k\rangle = \hat{\mathbf{U}}|a_k\rangle. \qquad (8.36)$$

8.12.1 *Unitary operators*

What do we know about the operator $\hat{\mathbf{U}}$? To begin with, in the absence of time-dependent influences acting on the measured system (and especially if the system is closed, which for the moment we will assume), $\hat{\mathbf{U}}$ may depend on the time difference $t_2 - t_1$ but not on t_1 or t_2 individually.

Suppose, next, that at t_1 the generic outcome $|w\rangle\langle w|$ has been obtained. Since the probabilities of the possible outcomes at t_2 add up to unity, we must have that

$$\sum_k |\langle b_k|\hat{\mathbf{U}}(t_2 - t_1)|w\rangle|^2 = 1. \qquad (8.37)$$

Omitting the time dependence of $\hat{\mathbf{U}}$, we can rewrite the left-hand side of Eq. (8.37) in a number of ways:

$$\sum_k |\langle b_k|\hat{\mathbf{U}}|w\rangle|^2 = \sum_k \langle \hat{\mathbf{U}}w|b_k\rangle\langle b_k|\hat{\mathbf{U}}w\rangle = \langle \hat{\mathbf{U}}w|\hat{\mathbf{U}}w\rangle = \langle w|\hat{\mathbf{U}}^\dagger\hat{\mathbf{U}}|w\rangle,$$

and it goes without saying to you should be checking the validity of each equality. Equation (8.37) thus implies that

$$\hat{\mathbf{U}}^\dagger\hat{\mathbf{U}} = \hat{\mathbf{I}}. \qquad (8.38)$$

This makes $\hat{\mathbf{U}}$ a *unitary operator*. By definition of the inverse operator $\hat{\mathbf{U}}^{-1}$, which undoes the action of $\hat{\mathbf{U}}$, we also have that

$$\hat{\mathbf{U}}^{-1}\hat{\mathbf{U}} = \hat{\mathbf{I}}. \qquad (8.39)$$

A unitary operator may therefore also be characterized by saying that its inverse equals its adjoint:

$$\hat{\mathbf{U}}^{-1} = \hat{\mathbf{U}}^\dagger. \qquad (8.40)$$

Problem 8.28. (∗) *The scalar product is invariant under unitary transformations:* $\langle \hat{\mathbf{U}}w|\hat{\mathbf{U}}v\rangle = \langle w|v\rangle$.

Problem 8.29. (∗) *The eigenvalues of unitary operators are phase factors.*

Problem 8.30. (∗) *If* $\hat{\mathbf{U}}|u\rangle = \lambda|u\rangle$ *then* $\hat{\mathbf{U}}^\dagger|u\rangle = \lambda^{-1}|u\rangle$.

Problem 8.31. (∗) *Eigenvectors of a unitary operator corresponding to distinct eigenvalues are orthogonal.*

Problem 8.32. $\left[\sum_{k=1}^{K} \lambda_k |a_k\rangle\langle a_k| \right]^n = \sum_{k=1}^{K} \lambda_k^n |a_k\rangle\langle a_k|$.

As you will recall, for every self-adjoint operator $\hat{\mathbf{A}}$ one can construct a basis made up of eigenvectors of $\hat{\mathbf{A}}$. The same is true for every unitary operator $\hat{\mathbf{U}}$. The spectral decomposition of a unitary operator has the form

$$\hat{\mathbf{U}} = \sum_k e^{i\alpha_k} |a_k\rangle\langle a_k|. \tag{8.41}$$

Let $\hat{\mathbf{A}}$ be self-adjoint. If we define $e^{i\hat{\mathbf{A}}}$ via the Taylor series of the exponential function and use the spectral decomposition of $\hat{\mathbf{A}}$, we find that the eigenvectors of $e^{i\hat{\mathbf{A}}}$ are the same as those of $\hat{\mathbf{A}}$. And if the corresponding eigenvalues of $\hat{\mathbf{A}}$ are λ_k, then the corresponding eigenvalues of $\hat{\mathbf{U}}$ are the phase factors $e^{i\lambda_k}$:

$$
\begin{aligned}
e^{i\hat{\mathbf{A}}} &= \sum_{n=0}^{\infty} \frac{(i\hat{\mathbf{A}})^n}{n!} \\
&= \sum_{n=0}^{\infty} \frac{i^n}{n!} \left[\sum_{k=1}^{K} \lambda_k |a_k\rangle\langle a_k| \right]^n \\
&= \sum_{n=0}^{\infty} \frac{i^n}{n!} \sum_{k=1}^{K} \lambda_k^n |a_k\rangle\langle a_k| \\
&= \sum_{n=0}^{\infty} \sum_{k=1}^{K} \frac{(i\lambda_k)^n}{n!} |a_k\rangle\langle a_k| \\
&= \sum_{k=1}^{K} e^{i\lambda_k} |a_k\rangle\langle a_k|.
\end{aligned} \tag{8.42}
$$

Thus every unitary operator $\hat{\mathbf{U}}$ can be cast into the form $e^{i\hat{\mathbf{A}}}$, where $\hat{\mathbf{A}}$ is self-adjoint.

There's something else we know about $\hat{\mathbf{U}}(t_2 - t_1)$. Since $(t_2 - t_1)$ is the amount of time we wait before performing the second measurement, and since waiting from t_1 to t_2 is the same as waiting from t_1 to $t' < t_2$ and then waiting some more, from t' to t_2, we have that

$$\hat{\mathbf{U}}(t_2 - t_1) = \hat{\mathbf{U}}(t_2 - t') \hat{\mathbf{U}}(t' - t_1), \tag{8.43}$$

and thus

$$e^{i\hat{\mathbf{A}}(t_2 - t_1)} = e^{i\hat{\mathbf{A}}(t_2 - t')} e^{i\hat{\mathbf{A}}(t' - t_1)}. \tag{8.44}$$

Since we also have that

$$e^{i\hat{\mathbf{A}}(t_2 - t')} e^{i\hat{\mathbf{A}}(t' - t_1)} = e^{i[\hat{\mathbf{A}}(t_2 - t') + \hat{\mathbf{A}}(t' - t_1)]},$$

we find that $\hat{\mathbf{A}}$ depends *linearly* on its argument:

$$\hat{\mathbf{A}}(t_2 - t') + \hat{\mathbf{A}}(t' - t_1) = \hat{\mathbf{A}}\big((t_2 - t') + (t' - t_1)\big) = \hat{\mathbf{A}}(t_2 - t_1).$$

This allows us to introduce a self-adjoint operator $\hat{\mathbf{H}}$, known as the *Hamilton operator* or simply the *Hamiltonian*, such that

$$\hat{\mathbf{U}}(\Delta t) = e^{-(i/\hbar)\hat{\mathbf{H}}\Delta t}. \tag{8.45}$$

The negative sign is again a convention. The division by $\hbar$ ensures that the eigenvalues of $\hat{\mathbf{H}}$ are measured in energy units. For infinitesimal dt we thus have that

$$\hat{\mathbf{U}}(dt) = \hat{\mathbf{I}} - \frac{i}{\hbar}\hat{\mathbf{H}}\,dt. \tag{8.46}$$

If we apply both sides to a vector $|v\rangle$ and use $|v(t + dt)\rangle = \hat{\mathbf{U}}(dt)|v(t)\rangle$, we obtain

$$|v(t + dt)\rangle - |v(t)\rangle = -\frac{i}{\hbar}\hat{\mathbf{H}}|v(t)\rangle\,dt, \tag{8.47}$$

and thus

$$\frac{d|v\rangle}{dt} = -\frac{i}{\hbar}\hat{\mathbf{H}}|v\rangle. \tag{8.48}$$

If the experimental conditions are time-dependent, $\hat{\mathbf{U}}$ depends on t in addition to its dependence on dt. This additional dependence goes into $\hat{\mathbf{H}}$.

If we use vector components $v_j = \langle a_j|v\rangle$ and operator components $H_{jk} = \langle a_j|\hat{\mathbf{H}}|a_k\rangle$, we can cast Eq. (8.48) into the form

$$i\hbar\frac{dv_j}{dt} = \sum_{k=1}^{K} H_{jk}v_k \quad \text{(for } j = 1, \ldots, K\text{)}. \tag{8.49}$$

8.12.2 Continuous variables

When dealing with a continuous variable—say, the position of a particle in one spatial dimension—it is customary to introduce a non-denumerably infinite basis of "improper" vectors $|x\rangle$. With respect to such a basis, a generic vector $|v\rangle$ has the components $v(x) = \langle x|v\rangle$, and the Hamiltonian has the components $H(x, x') = \langle x|\hat{\mathbf{H}}|x'\rangle$. In place of the sum $|v\rangle = \sum_k |a_k\rangle\langle a_k|v\rangle$ we then have the integral

$$|v\rangle = \int_{-\infty}^{+\infty} dx\, |x\rangle\langle x|v\rangle, \qquad (8.50)$$

and in place of Eq. (8.49) we have

$$i\hbar\frac{\partial v(x, t)}{\partial t} = \int_{-\infty}^{+\infty} dx'\, H(x, x')\, v(x', t). \qquad (8.51)$$

In addition, the orthonormality condition $\langle a_i|a_k\rangle = \delta_{ik}$ (Eq. 8.9) gets replaced by $\langle x|x'\rangle = \delta(x - x')$. The delta distribution $\delta(x - x')$ is defined by requiring that for any continuous function $f(x)$,

$$\int_{x_1}^{x_2} dx\, \delta(x - x_0)f(x) = f(x_0), \qquad (8.52)$$

where $x_1 < x_0 < x_2$.[5]

As it stands, however, Eq. (8.51) is inconsistent with the special theory of relativity (Chap. 6). The problem is that the left-hand side depends on the time rate of change of v for a *single* value of x, while the right-hand side depends on the simultaneous value of v for *every* value of x. The instantaneous action at a distance implied by this dependence is not possible in a relativistic world. The remedy is to include a delta distribution in the definition of the Hamiltonian: $H(x, x') = \delta(x - x')\,\hat{H}$. Because $H(x, x')$ may depend on how $v(x, t)$ changes *locally*, across infinitesimal intervals, $\hat{H}$ may contain differential operators with respect to x. Hence the hat: $\hat{H}$ is still an operator, albeit one acting on the function $v(x, t)$ rather than one acting on the vector $|v\rangle$. Equation (8.51) now takes the acceptable form

$$i\hbar\frac{\partial v(x, t)}{\partial t} = \hat{H}\, v(x, t). \qquad (8.53)$$

Problem 8.33. *Write down the Hamiltonians for Eq. (4.4) and Eq. (7.21).*

---

[5]For further discussion of the delta distribution see Marchildon (2002, Sec. 5.9.1) or Peres (1995, Sec. 4.7).

8.13　The rules of the game derived at last

The origin of the two rules formulated in Sec. 5.1, one of which led us to the Schrödinger equation via the Feynman route, is now readily understood.

Suppose that a maximal test performed at the time t_1 yields the outcome $|u\rangle\langle u|$, or u for short, and that we want to calculate the probability with which a maximal test performed at the time $t_2 > t_1$ yields the outcome w. Assume that at some intermediate time t another maximal test is made, and that its possible outcomes are v_i ($i = 1,\ldots,n$). Because a maximal test "prepares from scratch," as we have seen in Sec. 8.11, the joint probability $p(w, v_i|u)$ with which the intermediate and final tests yield v_i and w, respectively, given the initial outcome u, is the product of two probabilities: the probability $p(v_i|u)$ of v_i given u, and the probability $p(w|v_i)$ of w given v_i. By Born's rule (8.32), this is

$$p(w, v_i|u) = |\langle w|v_i\rangle\langle v_i|u\rangle|^2. \tag{8.54}$$

The probability of w given u, *regardless* of the intermediate outcome, is given by Rule A: first take the absolute squares of the amplitudes $A_i = \langle w|v_i\rangle\langle v_i|u\rangle$, then add the results:

$$p_A(w|u) = \sum_i p(w, v_i|u) = \sum_i |\langle w|v_i\rangle\langle v_i|u\rangle|^2. \tag{8.55}$$

If *no* intermediate measurement is made, we have $p_B(w|u) = |\langle w|u\rangle|^2$, and if we insert the identity operator $\hat{I} = \sum_i |v_i\rangle\langle v_i|$, we obtain

$$p_B(w|u) = \left|\sum_i \langle w|v_i\rangle\langle v_i|u\rangle\right|^2. \tag{8.56}$$

In other words, first add the amplitudes A_i, then take the absolute square of the result.

Chapter 9

The classical forces: Effects

9.1 The principle of "least" action

Let us return to the propagator for a stable scalar particle (Eq. 7.13):

$$\langle \mathbf{r}_B, t_B | \mathbf{r}_A, t_A \rangle = \int \mathcal{DC} \, e^{(i/\hbar) \int_C dS}. \tag{9.1}$$

The path integral sums contributions from all timelike paths that start at $\mathbf{r}_A$ at the time t_A and end at $\mathbf{r}_B$ at the time t_B. Why only timelike paths? Actually there is no law that excludes other paths. Such paths exclude themselves, as it were.

To see how, consider again a free particle, and suppose that $\mathcal{C}$ has a spacelike segment $\Delta\mathcal{C}$. In this case we have that $ds^2 < 0$ for all infinitesimal segments $d\mathcal{C}$ that make up $\Delta\mathcal{C}$. The integral $\int_{\Delta\mathcal{C}} ds$ is therefore imaginary. This means that the amplitude $e^{(i/\hbar) \int_C ds}$ contains the factor $e^{(i/\hbar) \int_{\Delta C} ds}$, which has a real exponent. This factor (and hence the amplitude) either blows up or falls off exponentially with (roughly) the distance between the endpoints of $\Delta\mathcal{C}$. If the propagator is to yield finite probabilities, it must fall off exponentially. As a result, spacelike ("superluminal") propagation is exponentially suppressed. It is not impossible, but it is very unlikely, except over very short distances.

What then is the probability of finding that the particle has traveled from spacetime point $(\mathbf{r}_A, t_A)$ to spacetime point $(\mathbf{r}_B, t_B)$ via a specific path $\mathcal{C}_0$? Since the magnitude of $Z[\mathcal{C}] = e^{(i/\hbar) \int_C ds}$ is the same for all paths from $(\mathbf{r}_A, t_A)$ to $(\mathbf{r}_B, t_B)$, this probability seems to be the same for all paths. It is, however, strictly impossible to ascertain, by any kind of measurement or sequence of measurements, that a particle has traveled via a definite path. We obtain a more useful answer if we make the half realistic assumption that what *can* be ascertained is whether a particle has traveled from $(\mathbf{r}_A, t_A)$ to $(\mathbf{r}_B, t_B)$ inside a narrow *bundle* of paths.

Imagine a narrow "tube" T filled with paths from $(\mathbf{r}_A, t_A)$ to $(\mathbf{r}_B, t_B)$. The probability of finding that the particle has traveled from $(\mathbf{r}_A, t_A)$ to $(\mathbf{r}_B, t_B)$ inside T is (formally) given by the absolute square of the path integral

$$I_T = \int_T \mathcal{DC} \, e^{(i/\hbar)S[\mathcal{C}]}, \qquad (9.2)$$

which sums over the paths from $(\mathbf{r}_A, t_A)$ to $(\mathbf{r}_B, t_B)$ that are contained in T. We shall assume that there is a single path from $(\mathbf{r}_A, t_A)$ to $(\mathbf{r}_B, t_B)$ for which the action $S[\mathcal{C}]$ is *stationary*: it does not change under small variations of this path.[1] Since the action differential dS defines a spacetime geometry (Sec. 7.1), this is the same as assuming that there is a single geodesic $\mathcal{G}$ connecting $(\mathbf{r}_A, t_A)$ with $(\mathbf{r}_B, t_B)$. (A *geodesic* is a path $\mathcal{G}$ that is either longer or shorter than all paths that have the same endpoints as $\mathcal{G}$ and lie in a sufficiently small neighborhood of $\mathcal{G}$.)

If we further assume that $\mathcal{G}$ lies inside T, then there is a neighborhood $\mathcal{N}$ of $\mathcal{G}$ inside T such that the phases $S[\mathcal{C}]/\hbar$ of the paths contained in $\mathcal{N}$ are almost equal to that of $\mathcal{G}$. As a result, the sum of the corresponding phase factors $e^{(i/\hbar)S[\mathcal{C}]}$ is of considerable magnitude, and so is the path integral I_T. This is illustrated by the almost straight chain of arrows in Fig. 9.1, which represent the almost equal phase factors contributed by the paths inside $\mathcal{N}$.

If $\mathcal{G}$ does not lie inside T, and if the differences between the phases of the paths contained in T are sufficiently large, the magnitude of the sum of the corresponding phase factors $e^{(i/\hbar)S[\mathcal{C}]}$ will be minute by comparison. This is illustrated by the coiled chain of arrows in Fig. 9.1.

How large is "sufficiently large"?

For a free and stable particle, whose action is given by $S[\mathcal{C}] = -\int mc^2 ds$ (Eq. 7.5), "sufficiently large" translates to "sufficiently large mass." In the limit $m \to \infty$, the only paths that contribute to I_T are those in the infinitesimal neighborhood of $\mathcal{G}$.

For a stable particle moving under the kind of influence that is represented by the potentials V and $\mathbf{A}$ (Eq. 7.8), the relevant theoretical limit is the classical limit $\hbar \to 0$. In this limit, it is safe to assume that a particle traveling from $(\mathbf{r}_A, t_A)$ to $(\mathbf{r}_B, t_B)$ travels via $\mathcal{G}$, as classical physics asserts. A classical particle follows a geodesic of the geometry defined by the action differential dS. If dS is given by Eq. (7.8), the resulting differential geometry is of a type known as *Finsler geometry* (Antonelli *et al.*, 1993; Rund, 1969).

---

[1] This assumption ensures that a unique path is obtained in the classical limit.

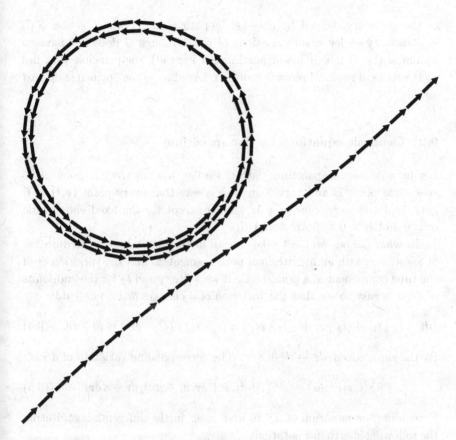

Fig. 9.1 The almost straight chain of arrows represents the almost equal phase factors contributed by the paths inside a narrow bundle that contains the geodesic $\mathcal{G}$. The coiled chain of arrows represents the phase factors contributed by the paths inside a narrow bundle that does not contain $\mathcal{G}$.

If we are dealing with a system that has N degrees of freedom, the *configuration* of the system is equivalent to a point in an N-dimensional *configuration space*, and the propagator sums contributions from all paths in the system's $N+1$-dimensional configuration space*time* $\mathcal{M}$ that lead from a point $\mathcal{P}_1$ to a point $\mathcal{P}_2$:

$$\langle \mathcal{P}_2, t_2 | \mathcal{P}_1, t_1 \rangle = \int \mathcal{DC}\, e^{(i/\hbar)S[\mathcal{C}]}. \qquad (9.3)$$

($\mathcal{P}_1$ specifies the values of all degrees of freedom at the time t_1, and $\mathcal{P}_2$ does the same for t_2.) In the classical limit, this system too follows a geodesic of the geometry defined by dS—i.e., a path $\mathcal{G}$ in $\mathcal{M}$ whose action $S[\mathcal{G}]$ is stationary under small variations of $\mathcal{G}$. Although $\mathcal{G}$ does not have to minimize the action (it might maximize it instead), the principle that lies at the roots of classical physics is often referred to as the "principle of *least* action."

9.2   Geodesic equations for flat spacetime

Let us now take a spacetime path $\mathcal{C}$ leading from spacetime point A to spacetime point B and vary it in such a way that every point $(\mathbf{r}, t)$ of $\mathcal{C}$ gets displaced to a point $(\mathbf{r} + \delta\mathbf{r}, t + \delta t)$ except for the fixed end points: $\delta t = 0$ and $\delta\mathbf{r} = 0$ at both A and B.

In what follows we must take care to distinguish between the duration dt associated with an infinitesimal path segment $d\mathcal{C}$ and the variation δt of the time component of a point on $\mathcal{C}$. If we write t_1 and t_2 for the endpoints of dt, it is easy to see that the variation of $\mathcal{C}$ changes dt into $dt + d\,\delta t$:

$$dt = t_2 - t_1 \rightarrow (t_2 + \delta t_2) - (t_1 + \delta t_1) = (t_2 - t_1) + (\delta t_2 - \delta t_1) = dt + d\,\delta t. \quad (9.4)$$

By the same token, $d\mathbf{r} \rightarrow d\mathbf{r} + d\,\delta\mathbf{r}$. The corresponding variation of dS is

$$dS(t, \mathbf{r}, dt, d\mathbf{r}) \rightarrow dS(t + \delta t, \mathbf{r} + \delta\mathbf{r}, dt + d\,\delta t, d\mathbf{r} + d\,\delta\mathbf{r}). \quad (9.5)$$

Expanding the variation of dS to first order in the differentials and using the following shorthand notations,

$$\frac{\partial\, dS}{\partial \mathbf{r}} \cdot \delta\mathbf{r} = \frac{\partial\, dS}{\partial x}\,\delta x + \frac{\partial\, dS}{\partial y}\,\delta y + \frac{\partial\, dS}{\partial z}\,\delta z\,,$$

$$\frac{\partial\, dS}{\partial\, d\mathbf{r}} \cdot d\,\delta\mathbf{r} = \frac{\partial\, dS}{\partial\, dx}\, d\,\delta x + \frac{\partial\, dS}{\partial\, dy}\, d\,\delta y + \frac{\partial\, dS}{\partial\, dz}\, d\,\delta z\,, \quad (9.6)$$

we obtain

$$dS(t, \mathbf{r}, dt, d\mathbf{r}) + \frac{\partial\, dS}{\partial t}\,\delta t + \frac{\partial\, dS}{\partial \mathbf{r}} \cdot \delta\mathbf{r} + \frac{\partial\, dS}{\partial\, dt}\, d\,\delta t + \frac{\partial\, dS}{\partial\, d\mathbf{r}} \cdot d\,\delta\mathbf{r}. \quad (9.7)$$

The difference between the varied and the original path elements, accordingly, is

$$\delta dS = \frac{\partial\, dS}{\partial t}\,\delta t + \frac{\partial\, dS}{\partial \mathbf{r}} \cdot \delta\mathbf{r} + \frac{\partial\, dS}{\partial\, dt}\, d\,\delta t + \frac{\partial\, dS}{\partial\, d\mathbf{r}} \cdot d\,\delta\mathbf{r}\,. \quad (9.8)$$

We now use the product rule (Eq. 3.13),

$$d\left[\frac{\partial\,dS}{\partial\,dt}\,\delta t\right] = \left[d\,\frac{\partial\,dS}{\partial\,dt}\right]\delta t + \frac{\partial\,dS}{\partial\,dt}\,d\,\delta t\,, \tag{9.9}$$

$$d\left[\frac{\partial\,dS}{\partial\,d\mathbf{r}}\cdot\delta\mathbf{r}\right] = \left[d\,\frac{\partial\,dS}{\partial\,d\mathbf{r}}\right]\cdot\delta\mathbf{r} + \frac{\partial\,dS}{\partial\,d\mathbf{r}}\cdot d\,\delta\mathbf{r}\,, \tag{9.10}$$

to replace the partial derivatives with respect to dt and $d\mathbf{r}$:

$$\delta\,dS = \frac{\partial\,dS}{\partial t}\,\delta t + \frac{\partial\,dS}{\partial\mathbf{r}}\cdot\delta\mathbf{r} + d\left[\frac{\partial\,dS}{\partial\,dt}\,\delta t\right] - \left[d\,\frac{\partial\,dS}{\partial\,dt}\right]\delta t$$

$$+ d\left[\frac{\partial\,dS}{\partial\,d\mathbf{r}}\cdot\delta\mathbf{r}\right] - \left[d\,\frac{\partial\,dS}{\partial\,d\mathbf{r}}\right]\cdot\delta\mathbf{r}\,. \tag{9.11}$$

Integrating over $\mathcal{C}$, we obtain the variation of $S[\mathcal{C}]$:

$$\delta S[\mathcal{C}] = \int_{\mathcal{C}}\left[\left(\frac{\partial\,dS}{\partial t} - d\,\frac{\partial\,dS}{\partial\,dt}\right)\delta t + \left(\frac{\partial\,dS}{\partial\mathbf{r}} - d\,\frac{\partial\,dS}{\partial\,d\mathbf{r}}\right)\cdot\delta\mathbf{r}\right]$$

$$+ \int_{\mathcal{C}}d\left[\frac{\partial\,dS}{\partial\,dt}\,\delta t + \frac{\partial\,dS}{\partial\,d\mathbf{r}}\cdot\delta\mathbf{r}\right]. \tag{9.12}$$

The second integral only depends on the endpoints of $\mathcal{C}$, at which $\delta t = 0$ and $\delta\mathbf{r} = 0$. It therefore vanishes. If $\mathcal{C}$ is a geodesic, the action is stationary, so that $\delta S[\mathcal{C}]$ vanishes, too. The first integral then vanishes as well. Moreover, since this vanishes for all possible variations δt and $\delta\mathbf{r}$ along $\mathcal{C}$, its integrand itself vanishes. The bottom line is that the geodesics defined by dS satisfy the *geodesic equations*

$$\frac{\partial\,dS}{\partial t} = d\,\frac{\partial\,dS}{\partial\,dt}\,, \tag{9.13}$$

$$\frac{\partial\,dS}{\partial\mathbf{r}} = d\,\frac{\partial\,dS}{\partial\,d\mathbf{r}}\,. \tag{9.14}$$

9.3   Energy and momentum

If we look at these equations, what jumps out at us right away is that if dS has no explicit time-dependence (i.e., $\partial\,dS/\partial t = 0$), then the system's *energy*

$$E \stackrel{\text{Def}}{=} -\frac{\partial\,dS}{\partial\,dt} \tag{9.15}$$

is constant along geodesics—the paths that classical systems follow in their configuration spacetimes. (We'll get to the reason for the negative sign in a moment.) Likewise, if dS has no explicit dependence on the components of $\mathbf{r}$ (i.e., $\partial\, dS/\partial\mathbf{r} = 0$), then the system's *momentum*

$$\mathbf{p} \overset{\text{Def}}{=} \frac{\partial\, dS}{\partial\, d\mathbf{r}} \tag{9.16}$$

is constant along geodesics.

E tells us how much the projection dt of a segment $d\mathcal{C}$ of a path $\mathcal{C}$ onto the time axis contributes to $S[\mathcal{C}]$, and $\mathbf{p}$ tells us how much the projection $d\mathbf{r}$ of $d\mathcal{C}$ onto space—the hyperplane spanned by the space axes—contributes to $S[\mathcal{C}]$. If dS has no explicit time dependence, then equal intervals of the time axis make equal contributions to $S[\mathcal{C}]$: they are *physically equivalent*: they represent *equal durations*. If dS has no explicit dependence on any space coordinate, then equal intervals of the space axes make equal contributions to $S[\mathcal{C}]$: they are *physically equivalent*: they represent *equal distances*.

The take-home message here is that energy is *defined* as the physical quantity whose constancy warrants the physical equivalence of equal time intervals, often called the "homogeneity of time," while momentum is *defined* as the physical quantity whose constancy warrants the physical equivalence of equal space intervals, often called the "homogeneity of space." If energy and/or momentum is/are not conserved, so that equal intervals of the time and/or space axis are not physically equivalent, the reason could be either that the system is not freely moving—an external influence renders quantitatively equal intervals physically inequivalent—or that we are using the wrong coordinates: coordinates that give rise to fictitious forces.

More can be discovered by differentiating Eq. (7.7) with respect to u. For the left-hand side we obtain

$$\frac{d(dS)}{du} = \frac{\partial\, dS}{\partial(u\, dt)}\frac{\partial(u\, dt)}{\partial u} + \frac{\partial\, dS}{\partial(u\, d\mathbf{r})}\cdot\frac{\partial(u\, d\mathbf{r})}{\partial u} = \frac{\partial\, dS}{\partial(u\, dt)}\, dt + \frac{\partial\, dS}{\partial(u\, d\mathbf{r})}\cdot d\mathbf{r}\,,$$

while the right-hand side comes out equal to dS. If we now set $u = 1$ and use definitions (9.15) and (9.16), we arrive at

$$-E\, dt + \mathbf{p}\cdot d\mathbf{r} = dS\,. \tag{9.17}$$

Since dS is a 4-scalar, the left-hand side, too, has to be a 4-scalar. Furthermore, since $(c\, dt, d\mathbf{r})$ are the components of a 4-vector, the left-hand side has to be the scalar product of two 4-vectors. The second 4-vector, $(E/c, \mathbf{p})$, is the particle's *energy–momentum* or *4-momentum*. The reason why we defined E and $\mathbf{p}$ with opposite signs should now be clear.

Problem 9.1. *If $ds = \sqrt{dt^2 - d\mathbf{r} \cdot d\mathbf{r}/c^2}$, then*

$$\frac{\partial\, ds}{\partial\, dt} = \frac{1}{\sqrt{1 - v^2/c^2}} \quad and \quad \frac{\partial\, ds}{\partial\, d\mathbf{r}} = -\frac{\mathbf{v}/c^2}{\sqrt{1 - v^2/c^2}}.$$

Plugging the action differential of Sec. 7.2,

$$dS = -mc^2\, ds - qV(t,\mathbf{r})\, dt + \frac{q}{c}\mathbf{A}(t,\mathbf{r}) \cdot d\mathbf{r}, \qquad (9.18)$$

into the definitions (9.15) and (9.16) of E and $\mathbf{p}$, we find that

$$E = \frac{mc^2}{\sqrt{1 - v^2/c^2}} + qV, \qquad (9.19)$$

$$\mathbf{p} = \frac{m\mathbf{v}}{\sqrt{1 - v^2/c^2}} + \frac{q}{c}\mathbf{A}. \qquad (9.20)$$

Both E and $\mathbf{p}$ are thus made up of a kinetic part and a potential part. The kinetic parts E_K and $\mathbf{p}_K$ are those containing the root; the potential parts are $E_P = qV$ and $\mathbf{p}_P = (q/c)\,\mathbf{A}$.

Problem 9.2. *The kinetic parts of (9.19) and (9.20) satisfy the relation*

$$E^2 - \mathbf{p}^2 c^2 = m^2 c^4. \qquad (9.21)$$

Problem 9.3. *Use the Taylor series to show that*

$$\left(1 - \frac{v^2}{c^2}\right)^{-1/2} = 1 + \frac{1}{2}\frac{v^2}{c^2} + \frac{3}{8}\frac{v^4}{c^4} + \cdots. \qquad (9.22)$$

Expanding the kinetic parts of (9.19) and (9.20) and dropping all terms having c in the denominator, we obtain the non-relativistic expressions

$$E = mc^2 + \frac{m}{2}v^2, \quad \mathbf{p} = m\mathbf{v}. \qquad (9.23)$$

Because the term mc^2 amounts to the presence of a (physically irrelevant) constant potential, it can be dropped.

9.4 Vector analysis: Some basic concepts

We already encountered the *gradient* of a function $f(x,y,z)$—a vector $\partial f/\partial\mathbf{r}$ whose components are

$$\frac{\partial f}{\partial x}, \frac{\partial f}{\partial y}, \frac{\partial f}{\partial z}. \qquad (9.24)$$

The "vector" $\partial/\partial \mathbf{r}$, whose components are the differential operators

$$\frac{\partial}{\partial x}, \frac{\partial}{\partial y}, \frac{\partial}{\partial z}, \tag{9.25}$$

is useful in a variety of ways. The product

$$d\mathbf{r} \cdot \frac{\partial}{\partial \mathbf{r}}, \tag{9.26}$$

for example, is a machine that accepts a function $f(x, y, z)$ and returns the difference between the respective values of f at the points $(x + dx, y + dy, z + dx)$ and (x, y, z):

$$df = \frac{\partial f}{\partial x} dx + \frac{\partial f}{\partial y} dy + \frac{\partial f}{\partial z} dz. \tag{9.27}$$

Problem 9.4. *The integral of a gradient along a curve $\mathcal{C}$ only depends on the endpoints of $\mathcal{C}$.*

9.4.1   Curl and Stokes's theorem

Another useful application of the operator $\partial/\partial \mathbf{r}$ is the *curl* of a vector field $\mathbf{A}(x, y, z)$, abbreviated to curl $\mathbf{A}$ and defined by

$$\frac{\partial}{\partial \mathbf{r}} \times \mathbf{A} = \left(\frac{\partial A_z}{\partial y} - \frac{\partial A_y}{\partial z} \right) \hat{\mathbf{x}} + \left(\frac{\partial A_x}{\partial z} - \frac{\partial A_z}{\partial x} \right) \hat{\mathbf{y}} + \left(\frac{\partial A_y}{\partial x} - \frac{\partial A_x}{\partial y} \right) \hat{\mathbf{z}}. \tag{9.28}$$

To uncover its significance, we calculate the *circulation* of $\mathbf{A}$ along a closed curve $\mathcal{C}$. This is given by the *line integral* $\oint \mathbf{A} \cdot d\mathbf{r}$ along $\mathcal{C}$. (A line integral along a closed curve is also called a *loop integral*.)

Let's start with the boundary of an infinitesimal rectangle with corners $A = (0, -dy/2, -dz/2)$, $B = (0, dy/2, -dz/2)$, $C = (0, dy/2, dz/2)$, and $D = (0, -dy/2, dz/2)$. The contributions from the four sides are, respectively,

$$\overline{AB}: \ +A_y(0, 0, -dz/2)\, dy = +\left(A_y(0,0,0) - \frac{\partial A_y}{\partial z} \frac{dz}{2} \right) dy,$$

$$\overline{BC}: \ +A_z(0, +dy/2, 0)\, dz = +\left(A_z(0,0,0) + \frac{\partial A_z}{\partial y} \frac{dy}{2} \right) dz,$$

$$\overline{CD}: \ -A_y(0, 0, +dz/2)\, dy = -\left(A_y(0,0,0) + \frac{\partial A_y}{\partial z} \frac{dz}{2} \right) dy,$$

$$\overline{DA}: \ -A_z(0, -dy/2, 0)\, dz = -\left(A_z(0,0,0) - \frac{\partial A_z}{\partial y} \frac{dy}{2} \right) dz. \tag{9.29}$$

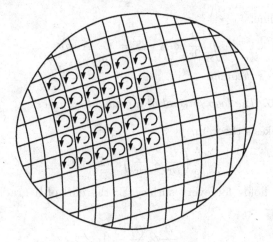

Fig. 9.2   Illustration of Stokes's theorem.

They add up to

$$\left(\frac{\partial A_z}{\partial y} - \frac{\partial A_y}{\partial z}\right) dy\,dz = \left(\frac{\partial}{\partial \mathbf{r}} \times \mathbf{A}\right)_x dy\,dz. \tag{9.30}$$

The next step is to represent our infinitesimal rectangle by a vector $d\mathbf{\Sigma}$ perpendicular to the rectangle and having a magnitude equal to the rectangle's area $dy\,dz$. This leaves us with two possible orientations. By convention, the orientation of $d\mathbf{\Sigma}$ is determined by another right hand rule: stick out your right thumb and curve the other fingers of your right hand. If the curved fingers indicate the direction of integration along the rectangle's boundary, the thumb indicates the orientation of $d\mathbf{\Sigma}$.

Thus we may write the circulation (9.30) as curl $\mathbf{A} \cdot d\mathbf{\Sigma}$. Now we take some finite surface Σ (Fig. 9.2) and divide it into infinitesimal rectangles like the one just discussed. (We cannot divide any old surface into *finite* rectangles, but keeping in mind the limit indicated by the word "infinitesimal," we can divide it into infinitesimal rectangles.) The sum of the circulations of all those infinitesimal rectangles is given by the surface integral

$$\int_{\Sigma} \text{curl } \mathbf{A} \cdot d\mathbf{\Sigma}.$$

A glance at Fig. 9.2 informs us that the adjacent sides of neighboring rectangles contribute with opposite signs, since they are integrated over in opposite directions. Their contributions to the surface integral thus cancel out and only the contributions from the boundary $\partial\Sigma$ of Σ survive.

The bottom line:

$$\int_\Sigma \frac{\partial}{\partial \mathbf{r}} \times \mathbf{A} \cdot d\Sigma = \oint_{\partial\Sigma} \mathbf{A} \cdot d\mathbf{r} \, . \tag{9.31}$$

This is *Stokes's theorem.*

Problem 9.5. *The loop integral of a gradient vanishes.*

Hence by Stokes's theorem,

$$\int_\Sigma \left(\frac{\partial}{\partial \mathbf{r}} \times \frac{\partial f}{\partial \mathbf{r}} \right) \cdot d\Sigma = 0 \, . \tag{9.32}$$

Because this holds for every surface Σ, the curl of a gradient vanishes identically:

$$\frac{\partial}{\partial \mathbf{r}} \times \frac{\partial f}{\partial \mathbf{r}} \equiv 0 \, . \tag{9.33}$$

9.4.2 Divergence and Gauss's theorem

Yet another useful application of the operator $\partial/\partial\mathbf{r}$ is the *divergence* of a vector field $\mathbf{B}(x, y, z)$, abbreviated to div $\mathbf{B}$ and defined by

$$\frac{\partial}{\partial \mathbf{r}} \cdot \mathbf{B} = \frac{\partial B_x}{\partial x} + \frac{\partial B_y}{\partial y} + \frac{\partial B_z}{\partial z} \, . \tag{9.34}$$

To uncover its significance, we take an infinitesimal rectangular cuboid of volume dx, dy, dz and calculate the net outward flux of $\mathbf{B}$ through its surface. The *flux* of a vector field $\mathbf{B}$ through a surface element $d\Sigma$ is given by $\mathbf{B} \cdot d\Sigma$.

There are three pairs of opposite sides. The net flux through the sides perpendicular to the x axis is

$$B_x(x + dx, y, z) \, dy \, dz - B_x(x, y, z) \, dy \, dz = \frac{\partial B_x}{\partial x} \, dx \, dy \, dz \, . \tag{9.35}$$

Ditto for the remaining pairs of sides. The net flux of $\mathbf{B}$ out of d^3r therefore equals

$$\left(\frac{\partial B_x}{\partial x} + \frac{\partial B_y}{\partial y} + \frac{\partial B_z}{\partial z} \right) dx \, dy \, dz = \frac{\partial}{\partial \mathbf{r}} \cdot \mathbf{B} \, d^3r \, . \tag{9.36}$$

Now we take some finite region R of space and divide it into infinitesimal rectangular cuboids like the one just discussed. The sum of the fluxes out of all those cuboids is given by the volume integral $\int_R$ div $\mathbf{B} \, d^3r$. Notice that the common sides of each pair of neighboring cuboids contribute twice with opposite signs—the flux out of one equals the flux into the other. This means that their contributions to the volume integral cancel out and only the contributions from the boundary ∂R of R survive. The bottom line:

$$\int_R \frac{\partial}{\partial \mathbf{r}} \cdot \mathbf{B} \, d^3 r = \int_{\partial R} \mathbf{B} \cdot d\mathbf{\Sigma} \,. \tag{9.37}$$

This is *Gauss's law*.

In the special case that $\mathbf{B}$ is the curl of a vector field $\mathbf{A}$, Gauss's law and Stokes's theorem imply that

$$\int_R \frac{\partial}{\partial \mathbf{r}} \cdot \frac{\partial}{\partial \mathbf{r}} \times \mathbf{A} \, d^3 r = \int_{\partial R} \frac{\partial}{\partial \mathbf{r}} \times \mathbf{A} \cdot d\mathbf{\Sigma} = \oint_{\partial \partial R} \mathbf{A} \cdot d\mathbf{r} \,. \tag{9.38}$$

On the right-hand side we integrate over the boundary $\partial \partial R$ of the boundary ∂R of R. But a boundary—in this case the closed surface ∂R—has no boundary. The loop integral therefore vanishes, and we end up with

$$\int_R \frac{\partial}{\partial \mathbf{r}} \cdot \frac{\partial}{\partial \mathbf{r}} \times \mathbf{A} \, d^3 r = 0 \,. \tag{9.39}$$

Because this holds for every region R, the divergence of a curl vanishes identically:

$$\frac{\partial}{\partial \mathbf{r}} \cdot \frac{\partial}{\partial \mathbf{r}} \times \mathbf{A} \equiv 0 \,. \tag{9.40}$$

Here are two useful identities involving both the curl and the divergence:

$$d\mathbf{r} \times \left(\frac{\partial}{\partial \mathbf{r}} \times \mathbf{A} \right) = \frac{\partial}{\partial \mathbf{r}} \left(\mathbf{A} \cdot d\mathbf{r} \right) - \left(d\mathbf{r} \cdot \frac{\partial}{\partial \mathbf{r}} \right) \mathbf{A} \,, \tag{9.41}$$

$$\frac{\partial}{\partial \mathbf{r}} \times \left(\frac{\partial}{\partial \mathbf{r}} \times \mathbf{A} \right) = \frac{\partial}{\partial \mathbf{r}} \left(\frac{\partial}{\partial \mathbf{r}} \cdot \mathbf{A} \right) - \left(\frac{\partial}{\partial \mathbf{r}} \cdot \frac{\partial}{\partial \mathbf{r}} \right) \mathbf{A} \,. \tag{9.42}$$

9.5   The Lorentz force

Equation (9.20) allows us to write the geodesic equation (9.14) in this form:

$$d\mathbf{p}_K + \frac{q}{c} d\mathbf{A} = \frac{\partial \, dS}{\partial \mathbf{r}} \,. \tag{9.43}$$

The left-hand side equals

$$d\mathbf{p}_K + \frac{q}{c} \left[\frac{\partial \mathbf{A}}{\partial t} \, dt + \left(d\mathbf{r} \cdot \frac{\partial}{\partial \mathbf{r}} \right) \mathbf{A} \right] \,. \tag{9.44}$$

With dS given by Eq. (9.18), the right-hand side equals

$$-q \frac{\partial V}{\partial \mathbf{r}} \, dt + \frac{q}{c} \frac{\partial}{\partial \mathbf{r}} \left(\mathbf{A} \cdot d\mathbf{r} \right) \,. \tag{9.45}$$

Making use of the identity (9.41), we cast this into the form

$$-q\frac{\partial V}{\partial \mathbf{r}}\,dt + \frac{q}{c}\left[\left(d\mathbf{r}\cdot\frac{\partial}{\partial \mathbf{r}}\right)\mathbf{A} + d\mathbf{r}\times\left(\frac{\partial}{\partial \mathbf{r}}\times\mathbf{A}\right)\right]. \qquad (9.46)$$

When we put both sides together, two terms cancel, and we end up with

$$d\mathbf{p}_K = q\left(-\frac{\partial V}{\partial \mathbf{r}} - \frac{1}{c}\frac{\partial \mathbf{A}}{\partial t}\right)dt + d\mathbf{r}\times\frac{q}{c}\left(\frac{\partial}{\partial \mathbf{r}}\times\mathbf{A}\right). \qquad (9.47)$$

The terms in brackets are known as the *electric field* $\mathbf{E}$ and the *magnetic field* $\mathbf{B}$, respectively:

$$\mathbf{E}\overset{\text{Def}}{=} -\frac{\partial V}{\partial \mathbf{r}} - \frac{1}{c}\frac{\partial \mathbf{A}}{\partial t}\,, \qquad \mathbf{B}\overset{\text{Def}}{=}\frac{\partial}{\partial \mathbf{r}}\times\mathbf{A}\,. \qquad (9.48)$$

Equation (9.47) thus reduces to

$$d\mathbf{p}_K = q\,\mathbf{E}\,dt + d\mathbf{r}\times\frac{q}{c}\,\mathbf{B}\,. \qquad (9.49)$$

A transparent result! As a classical charged particle travels along the segment $d\mathcal{G}$ of a geodesic of the geometry defined by the action differential (9.18), its kinetic momentum changes as described by two terms, one linear in the temporal component dt of $d\mathcal{G}$, the other linear in the spatial component $d\mathbf{r}$ of $d\mathcal{G}$. The change in the particle's kinetic momentum associated with the projection of $d\mathcal{G}$ onto the time axis is parallel to the electric field and proportional to its magnitude, while the change in $\mathbf{p}_K$ associated with the projection of $d\mathcal{G}$ onto space is perpendicular to both $d\mathbf{r}$ and the magnetic field (in compliance with the right hand rule) and proportional to the magnitude of the latter.

The transparency of this result is often obscured by dividing it by dt:[2]

$$\frac{d\mathbf{p}_K}{dt} = q\,\mathbf{E} + \frac{q}{c}\,\mathbf{v}\times\mathbf{B}\,. \qquad (9.50)$$

According to the *Lorentz force law*, as this form of Eq. (9.49) is known, the time-rate of change of $\mathbf{p}_K$ is the effect of two forces, an electric force $q\,\mathbf{E}$ and a magnetic force $(q/c)\,\mathbf{v}\times\mathbf{B}$.

It certainly is not easy to free ourselves from the primitive notion of force, which we derive from certain bodily sensations, like those associated with pushing and pulling. It ought to be clear, though, that such a notion can have little to do with the physical concept of force. When we look at the Lorentz force law, what we see is an equation by which we can calculate the time rate of change of the kinetic momentum of a charged particle

---

[2]Reminder: we are using the Gaussian system of units. In SI units the factor $1/c$ is absorbed into the definition of $\mathbf{A}$.

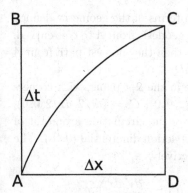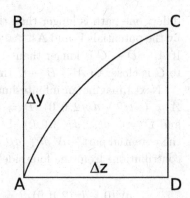

Fig. 9.3   If the action of the path $A \to B \to C$ is less than that of the path $A \to D \to C$, the geodesic from A to C—the path with the least action—is curved as indicated. Note that curvature in a spacetime plane amounts to acceleration or deceleration.

given the values of the computational tools **E** and **B**. What we do not see is anything justifying the interpretation of the right-hand side as a physical *agent* that *causes* the particle's kinetic momentum to change. In the next chapter we will look for the causes of the effects discussed in the present chapter. What we will not find is any physical mechanism or process by which these causes produce their effect. In particular, we will not find any justification for the classical story according to which the electromagnetic field physically implements the causal links between these causes and their effects.

9.5.1   *How the electromagnetic field bends geodesics*

A simple illustration of how the electromagnetic field relates to the (Finsler) geodesics associated with a charged particle should make it clear that we are dealing with a calculational tool rather than a physical agent.

Imagine a finite rectangle Q in the y–z plane, like that on the right of Fig. 9.3. By Stokes's theorem (Eq. 9.31), the magnetic flux through Q, $\int_Q \mathbf{B} \cdot d\mathbf{\Sigma}$, equals the circulation $\oint_{\partial Q} \mathbf{A} \cdot d\mathbf{r}$ of **A** around Q. This circulation is proportional to the action associated with the closed path $A \to B \to C \to D \to A$, which can also be interpreted as the difference between the respective actions associated with the paths $A \to B \to C$ and $A \to D \to C$.

If there is no electromagnetic field (in the equations), the two actions are equal, and the geodesic from A to C is a straight line. If there is a magnetic field (i.e., if the circulation of **A** around Q does not vanish), the two actions differ: one path is longer than the other (in terms of the geometry defined by the potentials V and $\mathbf{A}$). As a result, the geodesic from A to C is curved. If $A \to D \to C$ is longer than $A \to B \to C$, then the shortest path from A to C is closer to $A \to B \to C$ than to $A \to D \to C$.

Next, imagine an infinitesimal rectangle in the x–t plane, with corners $A = (-t/2, -dx/2, 0, 0)$, $B = (dt/2, -dx/2, 0, 0)$, $C = (dt/2, dx/2, 0, 0)$, and $D = (-dt/2, dx/2, 0, 0)$. Let us calculate the circulation around it of the potential part $-qV\,dt + (q/c)\mathbf{A} \cdot d\mathbf{r}$ of the action differential (9.18). The contributions from the four sides are, respectively,

$$\overline{AB}: \; -qV(0, -dx/2, 0, 0)\,dt = -q\left(V(0,0,0,0) - \frac{\partial V}{\partial x}\frac{dx}{2}\right)dt\,,$$

$$\overline{BC}: \; (q/c)A_x(dt/2, 0, 0, 0)\,dx = (q/c)\left(A_x(0,0,0,0) + \frac{\partial A_x}{\partial t}\frac{dt}{2}\right)dx\,,$$

$$\overline{CD}: \; qV(0, dx/2, 0, 0)\,dt = q\left(V(0,0,0,0) + \frac{\partial V}{\partial x}\frac{dx}{2}\right)dt\,,$$

$$\overline{DA}: \; -(q/c)A_x(-dt/2, 0, 0, 0)\,dx = -(q/c)\left(A_x(0,0,0,0) - \frac{\partial A_x}{\partial t}\frac{dt}{2}\right)dx\,,$$

and they add up to

$$q\left(\frac{\partial V}{\partial x} + \frac{1}{c}\frac{\partial A_x}{\partial t}\right)dt\,dx = -qE_x\,dt\,dx\,. \tag{9.51}$$

If we now add the circulations associated with the infinitesimal rectangles that make up a finite rectangle Q, like that on the left of Fig. 9.3, we again find that contributions from adjacent sides cancel (cf. Sec. 9.4.1). As a result, the circulation of the action around Q is given by the integral $-q \int_Q E_x(t, \mathbf{r})\,dt\,dx$.

This circulation can again be interpreted as the difference between the respective actions associated with the paths $A \to B \to C$ and $A \to D \to C$. If there is no electromagnetic field (in the equations), these actions are equal, and the geodesic from A to C is a straight line. Because we are now dealing with a spacetime plane containing the time axis, this means that the particle travels with a constant speed. If there is an electric field (i.e., if the circulation of the action around Q does not vanish), the two actions differ: one path is longer than the other (again, in terms of the geometry defined by V and $\mathbf{A}$). As a result, the geodesic from A to C is curved. In a plane containing the time axis, this means that the particle's speed changes; the particle is accelerating or decelerating.

9.6 Curved spacetime

So far in this chapter we have been concerned with the most straightforward way of formulating effects on the motion of a (scalar) particle, regardless of their causes. This consisted in adding to the action differential for a free particle a term that is linear in both dt and $d\mathbf{r}$, and it led to the differential geometry defined by (9.18). The latter describes all so-called electromagnetic effects on the motion of a (scalar) particle, both quantum-mechanically, via propagators, and classically, in terms of the curvature of geodesics.

There is one more way of incorporating effects on the motion of a (scalar) particle. It is to unfix the spacetime geometry implicit in the action differential for a free particle, $dS = -mc^2\, ds$. To implement this possibility, we begin by recasting the scalar product (6.53) of two 4-vectors into the following form:

$$(\vec{a}, \vec{b}) = \sum_{i=0}^{3} \sum_{k=0}^{3} g_{ik}\, a^i b^k. \tag{9.52}$$

The vector components are labeled by superscript indices running from 0 to 3—for example, $(a_t, a_x, a_y, a_z) = (a^0, a^1, a^2, a^3)$—and the numbers g_{ik} are as follows:

$$\begin{pmatrix} g_{00} & g_{01} & g_{02} & g_{03} \\ g_{10} & g_{11} & g_{12} & g_{13} \\ g_{20} & g_{21} & g_{22} & g_{23} \\ g_{30} & g_{31} & g_{32} & g_{33} \end{pmatrix} = \begin{pmatrix} 1 & 0 & 0 & 0 \\ 0 & -1 & 0 & 0 \\ 0 & 0 & -1 & 0 \\ 0 & 0 & 0 & -1 \end{pmatrix}. \tag{9.53}$$

When liberated from the specific values on the right-hand side, these numbers form the components of a pseudo-Riemannian *metric*—"Riemannian" because this metric is at the heart of the differential geometry originally formulated by Bernhard Riemann, and "pseudo" because even its "flat" form (given by the right-hand side of Eq. 9.53) is non-Euclidean, owing to the different signs of the diagonal terms. Because the scalar product (9.52) of two 4-vectors is symmetric, $(\vec{a}, \vec{b}) = (\vec{b}, \vec{a})$, we can require that the metric components be symmetric, too: $g_{ik} = g_{ki}$.

The metric is a machine with two input slots. If we insert into both slots the infinitesimal 4-vector with components $(dx^0, dx^1, dx^2, dx^3) = (c\, dt, dx, dy, dz)$, we obtain (recalling Eq. 6.45) the 4-scalar

$$\sum_{i=0}^{3} \sum_{k=0}^{3} g_{ik}\, dx^i\, dx^k = c^2\, dt^2 - d\mathbf{r} \cdot d\mathbf{r} = c^2\, ds^2. \tag{9.54}$$

If we adopt the convention (due to Einstein) that equal spacetime indices occurring twice (once as subscript and once as superscript) are summed automatically, without having to write those bulky summation symbols, we can substitute $g_{ik} \, dx^i \, dx^k$ for $c^2 \, ds^2$, and we can cast the action differential for a particle subject to such effects as are represented by a pseudo-Riemannian geometry, into the form

$$dS = -mc\sqrt{g_{ik} \, dx^i \, dx^k} \, . \tag{9.55}$$

9.6.1 *Geodesic equations for curved spacetime*

By allowing the metric components g_{ik} to differ from spacetime point to spacetime point, we make room for another kind of influence on the motion of a particle, regardless of what is exerting it. Its effects will again be mathematically expressed as modifications of the actions associated with spacetime paths, and classical particles will again follow the geodesics of the corresponding differential geometry.

The geodesic equations for the geometry defined by (9.55) can be obtained in much the same way we obtained the geodesic equations (9.13) and (9.14) for the geometry defined by the action differential (9.18). [For the details, the reader may consult Landau and Lifshitz (1975).] Here is the result:

$$\frac{d^2 x^i}{ds^2} + \Gamma^i_{kl} \frac{dx^k}{ds} \frac{dx^l}{ds} = 0 \, . \tag{9.56}$$

The symbols

$$\Gamma^i_{kl} \overset{\text{Def}}{=} \frac{1}{2} g^{im} \left(\frac{\partial g_{mk}}{\partial x^l} + \frac{\partial g_{ml}}{\partial x^k} - \frac{\partial g_{kl}}{\partial x^m} \right) \tag{9.57}$$

are the so-called connection coefficients, also known as "Christoffel symbols." (Superscript indices in denominators are to be treated like subscript indices in numerators.)

9.6.2 *Raising and lowering indices*

Before we can enter into the physical meaning of the geodesic equations (9.56), we need to know how *covariant* components (having subscript indices) are transformed into *contravariant* components (having superscript indices), and *vice versa*. The covariant components of a 4-vector are defined by

$$a_i \overset{\text{Def}}{=} g_{ik} \, a^k \, . \tag{9.58}$$

This tells us how to lower indices. Identical indices on different sides of an equation can be raised or lowered simultaneously. Thus $a^i = g^i_k a^k$. This tells us that

$$g^i_k = \begin{cases} 0 & \text{if } i \neq k \\ 1 & \text{if } i = k \end{cases}.$$ (9.59)

(Because of the symmetry of the metric, g^i_k can be read either as $g^i{}_k$ or as $g_k{}^i$.) The definition (9.58) allows us to cast Eq. (9.52) into the simple form

$$(\vec{a}, \vec{b}) = a_k b^k.$$ (9.60)

By the symmetry of the scalar product, $a_k b^k = a^k b_k$. This holds for any matching pair of indices: the covariant index can be raised if the contravariant index is lowered at the same time. Thus we also have that

$$a^i = g^{ik} a_k.$$ (9.61)

This tells us how to raise indices.

9.6.3 *Curvature*

The curvature of a 2-dimensional surface (such as the surface of the sphere in Fig. 9.4) is readily detected by us since we have the luxury of a consciousness capable of visualizing 3-dimensional objects. But how do we detect the curvature of a 3-dimensional hypersurface if we do not have the luxury of a consciousness capable of visualizing 4-dimensional objects? Or how do we detect the curvature of a 3-dimensional structure if this is not embedded in a 4-dimensional space? And, harder still, how do we detect the curvature of 4-dimensional spacetime?

Put yourself in the (2-dimensional) shoes of a flatlander.[3] You live in a 2-dimensional world and cannot leave it; you cannot imagine a third dimension perpendicular to your own two. How can you find out whether your world is curved?

Here is a possible test: draw a circle and measure its circumference C; draw a diameter of the circle and measure its length D; take the ratio C/D. If it differs from π, your world is curved. If it is smaller than π, you may live on the surface of a sphere (Fig. 9.4). If it is larger than π, you may live on a saddle-shaped surface.

---

[3] *Flatland: A Romance of Many Dimensions* is an 1884 satirical novella by the English schoolmaster Edwin A. Abbott. In a foreword to one of its many publications, Isaac Asimov wrote that it was "the best introduction one can find into the manner of perceiving dimensions."

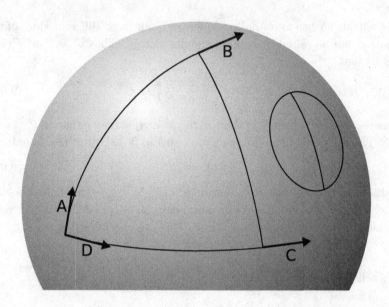

Fig. 9.4   Detecting the curvature of a surface without leaving it.

Another possible test: construct a triangle and measure its angles. If the sum of the angles differs from 180°, your world is curved. (Each of the angles of the triangle in Fig. 9.4 equals 90°, which makes a total of 270°.)

9.6.4   *Parallel transport*

Let us place the initial point ("tail") of a vector A at a corner of the triangle in Fig. 9.4, in such a way that A is tangent on one of the triangle's sides meeting at that corner. Let's call this side "S_1." Being straight, the entire vector—from its initial point to its terminal point ("head")—cannot lie on the sphere or inside its surface. It resides in a *tangent space* that we—but not the flatlanders—are able to visualize as a plane touching the sphere at a single point.

We want to transport A along S_1 in such a way that it remains parallel to itself. How do we do this? To find the answer, we should keep three things in mind:

(1) if we parallel-transport a vector V along a straight line $\mathcal{L}$, and if V is initially tangent on $\mathcal{L}$, V remains tangent on $\mathcal{L}$;

(2) in the presence of curvature, a geodesic is the next best thing to a straight line;

(3) the sides of a triangle on the surface of a sphere are geodesics. (This is part of what "triangle" means on the surface of a sphere.)

We shall accordingly make it part of what "parallel transport" means on the surface of a sphere that *geodesics parallel transport their tangent vectors.* For (2-dimensional) surfaces, a more general statement holds: if V is parallel transported along a geodesic $\mathcal{G}$, and if T is a co-moving tangent vector on $\mathcal{G}$ originating at the tail of V, the angle between V and $\mathcal{G}$ remains constant.

Thus when A (in Fig. 9.4) reaches the other end of S_1, it is equal to B; when subsequently parallel transported along the second side of the triangle, it becomes equal to C, and if it is then parallel transported along the third side, it becomes equal to D. This illustrates another manifestation of curvature: if a vector is parallel transported along a closed curve, so that it returns to its starting point *and* remains parallel to itself all along the way, it will yet, in general, point in a different direction than it did at the outset.

In 4-dimensional spacetime we can perform an analogous test. To find out if spacetime is warped, we may transport gyroscopes, initially pointing in the same direction, along different routes from a common starting point A to a common end point B. In a curved spacetime, they generally won't point in the same direction when they meet again at B.

To find out more, let $\vec{v}$ be a vector field in spacetime. We wish to know the difference $D\vec{v}$ between $\vec{v}(\mathcal{P})$ and $\vec{v}(\mathcal{P}')$ if the spacetime points $\mathcal{P}$ and $\mathcal{P}'$ are separated by an infinitesimal vector with components dx^k. Because the two vectors are situated in different tangent spaces—flat spacetimes "touching" curved spacetime at the respective points $\mathcal{P}$ and $\mathcal{P}'$—we need to take into account that the basis vectors of these tangent spaces may differ. If they are the same, the components of the two vectors simply differ by

$$Dv^i = dv^i = \frac{\partial v^i}{\partial x^l}\, dx^l. \tag{9.62}$$

If they are not the same, a compensatory term must be added. Because $\mathcal{P}$ and $\mathcal{P}'$ are an infinitesimal distance apart, this term will be linear in the components v^i and dx^k. Thus

$$Dv^i = dv^i + \overline{\Gamma}^i_{kl} v^k dx^l = \left(\frac{\partial v^i}{\partial x^l} + \overline{\Gamma}^i_{kl} v^k \right) dx^l. \tag{9.63}$$

Now suppose that $\mathcal{P}$ and $\mathcal{P}'$ are the endpoints of an infinitesimal segment $d\mathcal{G}$ of a geodesics $\mathcal{G}$, and that $\vec{u}$ is a tangent vector on $\mathcal{G}$ located at $\mathcal{P}$. Let us use a specific tangent vector, the *4-velocity* $u^i = dx^i/c\,ds$. (Here we conform to the convention of referring to a 4-vector $\vec{v}$ by a generic component v^i.)

Problem 9.6. $u^i = dx^i/c\,ds$ *satisfies* $u_i u^i = 1$.

Because geodesics parallel transport their tangent vectors, there will be no difference between $u^i(\mathcal{P})$ and $u^i(\mathcal{P}')$:

$$Du^i = du^i + \overline{\Gamma}^i_{kl} u^k dx^l = 0 \,. \tag{9.64}$$

Divide by ds to find that

$$\frac{d^2x^i}{ds^2} + \overline{\Gamma}^i_{kl}\frac{dx^k}{ds}\frac{dx^l}{ds} = 0 \,. \tag{9.65}$$

Both Eq. (9.56) and Eq. (9.65) characterizes the geodesics of the geometry defined by the action differential (9.55). The former defines them as the curves that minimize (or maximize) their actions, whereas the latter defines them as the curves that parallel transport their tangent vectors. Since the two equations are therefore identical, $\overline{\Gamma}^i_{kl}$ is given by Eq. (9.57)

A useful abbreviation,

$$v^i_{;l} \overset{\text{Def}}{=} \frac{\partial v^i}{\partial x^l} + \Gamma^i_{kl} v^k \,, \tag{9.66}$$

allows us to write $Dv^i = v^i_{;l}\,dx^l$ for the *covariant derivative* (9.63).

9.7 Gravity

While the geodesics defined by Eqs. (9.14) and (9.18) describe the classical effects of what history has led us to call "the electromagnetic force," the geodesics defined by Eq. (9.56) describe the classical effects of what history has led us to call "the gravitational force" or "gravity." One crucial difference between the corresponding geometries is that the former is particle-specific: it depends, through Eq. (9.18), on the masses and the charges of the particles affected. Equation (9.56), by contrast, is universal. It contains no particle-specific parameters, and for this reason it is customary to attribute the latter geometry to spacetime itself.

As we shall see in Part 3, quantum mechanics provides sufficient ground for rejecting the notion of an intrinsically differentiated and intrinsically structured spacetime. For the moment, however, I shall confine myself to illustrating that the mathematical description of gravitational effects in

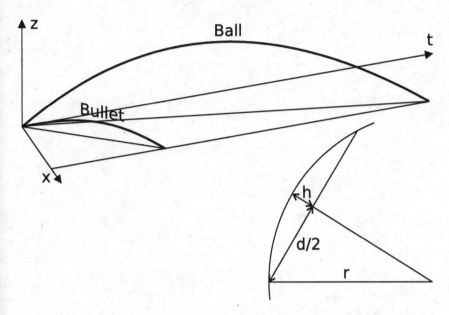

Fig. 9.5   The spacetime curvatures of the trajectories of a ball and of a bullet near the Earth's surface.

terms of a universal spacetime geometry is not as counterintuitive as it might seem at first. To this end we will estimate the spacetime curvatures of two trajectories observed near the Earth's surface:

- A ball is thrown so that it rises to a height of about $h_1 = 5$ m, covers a distance of 10 m, and hits the ground after 2 seconds.
- A bullet is fired so that it rises to a height of about $h_2 = 5 \times 10^{-4}$ m, covers the same distance, and hits the ground after 0.02 seconds.

In *space* the curvatures of these trajectories are obviously very different, but not in *spacetime*. The distances traveled in spacetime are essentially the times of travel multiplied by the speed of light, i.e., $d_1 = 2c = 6 \times 10^8$ m for the ball and $d_2 = 0.02\,c = 6 \times 10^6$ m for the bullet (Fig. 9.5). According to Pythagoras, the radius of curvature r is related to the parameters h and d via $(r - h)^2 + (d/2)^2 = r^2$. Because $h \ll d < r$, we ignore the term quadratic in h and obtain $r = d^2/8h$. This works out at $r = 9 \times 10^{15}$ m ≈ 1 light year for both the ball and the bullet.

Chapter 10

The classical forces: Causes

10.1 Gauge invariance

Problem 10.1. (∗) *Adding* $(q/c)\, df(t, \mathbf{r})$ *to the action differential* dS *induces the following transformation of the propagator (9.1):*

$$\langle \mathbf{r}_B, t_B | \mathbf{r}_A, t_A \rangle \to e^{(i/\hbar)(q/c)\, f(t_B, \mathbf{r}_B)} \langle \mathbf{r}_B, t_B | \mathbf{r}_A, t_A \rangle\, e^{-(i/\hbar)(q/c)\, f(t_A, \mathbf{r}_A)}.$$
(10.1)

It follows that Eq. (7.12),

$$\psi(\mathbf{r}_B, t_B) = \int d^3 r_A \, \langle \mathbf{r}_B, t_B | \mathbf{r}_A, t_A \rangle\, \psi(\mathbf{r}_A, t_A),$$

is invariant under the transformation (10.1), provided that this is combined with the phase transformation

$$\psi(t, \mathbf{r}) \to e^{(i/\hbar)(q/c)\, f(t, \mathbf{r})} \psi(t, \mathbf{r}).$$
(10.2)

Problem 10.2. (∗) *Adding* $(q/c)\, df(t, \mathbf{r})$ *to the action differential* dS *given by Eq. (7.8) is equivalent to the following transformation of the potentials:*

$$V \to V - \frac{1}{c} \frac{\partial f}{\partial t}, \qquad \mathbf{A} \to \mathbf{A} + \frac{\partial f}{\partial \mathbf{r}}.$$
(10.3)

In classical physics, one refers to (10.3) as a *gauge transformation*. In quantum physics, one usually understands by this term the combination of (10.3) with (10.2).

Problem 10.3. *The electric and magnetic fields (Eq. 9.48) are invariant under the transformation (10.3).*

10.2　Fuzzy potentials

The task now before us is to link to their causes the effects that we have just learned to describe in mathematical language. What, then, *are* their causes? Here we cannot proceed further without making specific assumptions. The causes *could* include minds, spirits, elves, goblins—what have you. Goblins appear to have been at work at the LHC, but otherwise we just don't know. We shall simply *assume* that we are dealing with particles and nothing but particles and that, therefore, the behavior of particles (including their aggregates) is affected by nothing but particles (including their aggregates). We shall however keep an open mind as to the nature of the things we call "particles" and "aggregates."

So far we have assumed that the potentials $V(t, \mathbf{r})$ and $\mathbf{A}(t, \mathbf{r})$ have exact values. Yet particles are *fuzzy*: neither their positions nor their momenta are in possession of exact values. If these potentials represent effects that particles have on particles, their values cannot be sharp. Fuzzy causes have fuzzy effects.

So how do we make room for fuzzy potentials? In much the same way that we made room for fuzzy particles! In the case of a single particle, this meant calculating the propagator $\langle \mathbf{r}_2, t_2 | \mathbf{r}_1, t_1 \rangle$ by summing over spacetime paths leading from $(\mathbf{r}_1, t_1)$ to $(\mathbf{r}_2, t_2)$. For a system with N degrees of freedom, this meant calculating the propagator $\langle \mathcal{P}_2, t_2 | \mathcal{P}_1, t_1 \rangle$ by summing over paths from a point $(\mathcal{P}_1, t_1)$ to a point $(\mathcal{P}_2, t_2)$ in the system's $N{+}1$-dimensional configuration spacetime (Eq. 9.3). Presently we are dealing with a mathematical device that allows us to calculate effects of a particular type. This device—the 4-potential $\vec{A} = (V, \mathbf{A})$—has a denumerably infinite number of degrees of freedom and a configuration space with as many dimensions. The configuration $\mathcal{A}$ obtaining at the time t specifies the components of $\vec{A}$ associated with each point $\mathbf{r}$ in the 3-dimensional hyperplane associated with t.[1]

To make room for fuzzy potentials, we thus calculate the propagator $\langle \mathcal{A}_2, t_2 | \mathcal{A}_1, t_1 \rangle$ by summing over all paths leading from a given configuration

---

[1] Like the real line or the complex plane, this hypersurface forms a *non-denumerable* set. How can the configuration of $\mathcal{A}$ be specified by a *denumerable* set of values? The answer is that we take $\vec{A}$ to be well-behaved in the sense spelled out in Sec. 3.3. Just as a well-behaved function of a single variable x can be fully specified by a countable set of parameters—for instance, the coefficients of its Taylor series—so can a well-behaved 4-vector field.

$\mathcal{A}_1$ (obtaining at the time t_1) to a given configuration $\mathcal{A}_2$ (obtaining at t_2):

$$\langle \mathcal{A}_2, t_2 | \mathcal{A}_1, t_1 \rangle = \int \mathcal{DC} \, e^{(i/\hbar) \, S_A[\mathcal{C}]}. \tag{10.4}$$

10.2.1 Lagrange function and Lagrange density

To find the appropriate action $S_A[\mathcal{C}]$, we begin by observing that the action differential $dS(t, \mathbf{r}, dt, d\mathbf{r})$ associated with an infinitesimal segment of a spacetime path can be cast into the form

$$dS = L(t, \mathbf{r}, \mathbf{v}) \, dt. \tag{10.5}$$

To see this, we only have to substitute $1/dt$ for the parameter u in Eq. (7.7). The immediate result is

$$dS(t, \mathbf{r}, \mathbf{v}) = \frac{dS(t, \mathbf{r}, dt, d\mathbf{r})}{dt}.$$

But now the notation is no longer appropriate, for the left-hand side, having ceased to depend on infinitesimal quantities, has itself ceased to be an infinitesimal quantity. In its place we write $L(t, \mathbf{r}, \mathbf{v})$. L is known as a *Lagrange function*.

Problem 10.4. (∗) *Write down the Lagrange function for the action differential (7.8).*

The homogeneity of dS expressed by Eq. (7.7) allowed us to interpret dS as defining a differential geometry. If we interpret dS_A as defining a differential geometry in the configuration spacetime of the potentials, then we also have that

$$dS_A = L_A(\mathcal{A}, d\mathcal{A}/dt) \, dt. \tag{10.6}$$

Consistency with special relativity, however, requires that the four dimensions of spacetime be treated on an equal footing (apart from the sign difference in Eq. 6.45). This means that S_A must be expressible as an integral over spacetime, rather than as an integral over time only:

$$S_A = \frac{1}{c} \int \mathcal{L}_A(A^k, \partial A^k / \partial x^i) \, d^4 x. \tag{10.7}$$

The *Lagrange density* $\mathcal{L}_A$ (also known simply as the *Lagrangian*) is a 4-scalar, and $d^4 x$ stands for the 4-volume element $(c \, dt) \, dx \, dy \, dz$.

In classical physics, the potentials serve to connect physical effects with their physical causes, but *only as calculational tools*. Given the causes, we calculate the potentials, and given the potentials, we calculate the effects.

Since the effects are represented by gauge-invariant combinations of the potentials (**E** and **B**), it makes sense to require that the causes determine only these combinations. We therefore require that $\mathcal{L}_\mathcal{A}$, too, be invariant under the transformation (10.2).[2] If we further require that the effects that charges have on charges are additive, in the sense that the combined effect of two charges A and B on a third charge is the (vectorial) sum of the individual effects produced by A and B, then we are left with a single candidate, the Lagrangian

$$\mathcal{L}_\mathcal{A} = \frac{1}{8\pi}(\mathbf{E}^2 - \mathbf{B}^2).$$
(10.8)

10.3   Maxwell's equations

As the fundamental equation of classical electrodynamics are formulated in terms of a current rather than in terms of particles, we replace the charge q in Eq. (9.18) by a charge density ρ and integrate over space as well:

$$S_{\mathcal{A}\mathcal{M}} = -\int (qV - \frac{q}{c}\mathbf{A}\cdot d\mathbf{v})\, dt \longrightarrow -\frac{1}{c^2}\int (c\rho V - \mathbf{A}\cdot \mathbf{j})\, d^4x.$$
(10.9)

The current density $\mathbf{j} = \rho\mathbf{v}$ denotes the flow of charge per unit area per unit time.

We have omitted $S_\mathcal{M} = -mc^2 ds$, the term in Eq. (9.18) that does not contain the potentials. When we established how the classical trajectories of charged particles depend on the potentials, we took the values of the potentials as given. We therefore had no need to include $S_\mathcal{A}$. Presently we wish to determine how **E** and **B** depend on the distribution and motion of particles (represented, respectively, by ρ and **j**). Since the distribution and motion of particles is now taken as given, we do not need to include $S_\mathcal{M}$.

To streamline our task, we define the tensor components

$$F_{ik} = \frac{\partial A_k}{\partial x^i} - \frac{\partial A_i}{\partial x^k},$$
(10.10)

the components A^k of $\vec{A}$ being given by (V, A_x, A_y, A_z). (Tensor components like F_{ik} transform like a product of vector components $C_i D_k$.)

---

[2]There is a deeper reason for this requirement: it makes quantum electrodynamics *renormalizable* (Sec. 15.8).

Problem 10.5.

$$\begin{pmatrix} F_{00} & F_{01} & F_{02} & F_{03} \\ F_{10} & F_{11} & F_{12} & F_{13} \\ F_{20} & F_{21} & F_{22} & F_{23} \\ F_{30} & F_{31} & F_{32} & F_{33} \end{pmatrix} = \begin{pmatrix} 0 & E_x & E_y & E_z \\ -E_x & 0 & -B_z & B_y \\ -E_y & B_z & 0 & -B_x \\ -E_z & -B_y & B_x & 0 \end{pmatrix}. \tag{10.11}$$

If in addition we introduce the 4-current $\vec{J}$ with components $J^k = (c\rho, j_x, j_y, j_z)$, we can write the relevant action more compactly as

$$S_{\mathcal{AM}} + S_{\mathcal{A}} = -\frac{1}{c^2} \int \left(A_k J^k + \frac{c}{16\pi} F_{ik} F^{ik} \right) d^4x. \tag{10.12}$$

Problem 10.6. *This is the same as*

$$-\frac{1}{c^2} \int (c\rho V - \mathbf{A} \cdot \mathbf{j}) \, d^4x + \frac{1}{8\pi c} \int (\mathbf{E}^2 - \mathbf{B}^2) \, d^4x. \tag{10.13}$$

If $\delta(S_{\mathcal{AM}} + S_{\mathcal{A}})$ is to vanish under any infinitesimal variation δA^k of the potentials, we must have that

$$\int \left[(\delta A_k) J^k + \frac{c}{8\pi} (\delta F_{ik}) F^{ik} \right] d^4x = 0. \tag{10.14}$$

Here we have made use of $\delta(F_{ik} F^{ik}) = 2(\delta F_{ik}) F^{ik}$, which holds for much the same reason as $dx^2 = 2x \, dx$. Because the antisymmetry of F_{ik} (i.e., $F_{ik} = -F_{ki}$) allows us to write

$$F^{ik}(\delta F_{ik}) = 2 F^{ik} \frac{\partial \delta A_k}{\partial x^i}, \tag{10.15}$$

Eq. (10.14) becomes

$$\int \left[(\delta A_k) J^k + \frac{c}{4\pi} F^{ik} \frac{\partial \delta A_k}{\partial x^i} \right] d^4x = 0. \tag{10.16}$$

Integrating by parts, we obtain

$$\int F^{ik} \frac{\partial \delta A_k}{\partial x^i} d^4x = \int \frac{\partial}{\partial x^i} \left(F^{ik} \delta A_k \right) d^4x - \int \frac{\partial F^{ik}}{\partial x^i} \delta A_k \, d^4x. \tag{10.17}$$

The 4-dimensional version of Gauss's law (9.37) allows us to convert the first integral on the right-hand side into an integral over a spatiotemporal boundary. At the spatial "boundary" (infinity) the potentials—at any rate, their gauge-invariant combinations—are assumed to vanish, and at the temporal boundaries the configurations $\mathcal{A}_1$ and $\mathcal{A}_2$ are fixed ($\delta\mathcal{A}_1 = \delta\mathcal{A}_2 = 0$). This integral therefore equals zero, so Eq. (10.14) takes the form

$$\int \left(J^k - \frac{c}{4\pi} \frac{\partial F^{ik}}{\partial x^i} \right) \delta A_k \, d^4x = 0. \tag{10.18}$$

Thus if $S_{\mathcal{AM}} + S_{\mathcal{A}}$ is to be stationary for every (infinitesimal) variation δA_k, the potentials must satisfy the condition

$$\frac{\partial F^{ik}}{\partial x^i} = \frac{4\pi}{c} J^k. \tag{10.19}$$

Problem 10.7. *Equation (10.19) is equivalent to*

$$\frac{\partial}{\partial \mathbf{r}} \cdot \mathbf{E} = 4\pi\rho, \qquad \frac{\partial}{\partial \mathbf{r}} \times \mathbf{B} = \frac{1}{c}\frac{\partial \mathbf{E}}{\partial t} + \frac{4\pi}{c}\mathbf{j}. \tag{10.20}$$

This is the second pair of *Maxwell's equations*. The first pair,

$$\frac{\partial}{\partial \mathbf{r}} \cdot \mathbf{B} = 0, \qquad \frac{\partial}{\partial \mathbf{r}} \times \mathbf{E} = -\frac{1}{c}\frac{\partial \mathbf{B}}{\partial t}, \tag{10.21}$$

follows from the definitions (9.48) of $\mathbf{E}$ and $\mathbf{B}$ and the identities (9.33) and (9.40). Together with the Lorentz force law (9.50), these four equations make up the set of fundamental equations of the classical electromagnetic theory.

The following problem illustrates that electricity and magnetism are two sides of the same coin: what counts as a magnetic effect if one reference frame is used, may count as an electric effect if a different frame is used, and *vice versa*.

Problem 10.8. (∗) *Consider a long current-bearing wire. Situated near the wire is a charged particle. Described in terms of one inertial frame (F_1), the particle is at rest. Described in terms of another inertial frame (F_2), the particle moves parallel to the wire with a constant speed. In F_2, the particle experiences a force according to the magnetic term of the Lorentz force law. In F_1, this term vanishes, so the force acting on the particle must be due to the electric term. What is the source of $\mathbf{E}$ in this frame?*

10.3.1   Charge conservation

Problem 10.9. (∗) *Use Eq. (10.19) to show that*

$$\frac{\partial J^k}{\partial x^k} = 0. \tag{10.22}$$

In 3-vector notation this reads:

$$\frac{\partial \rho}{\partial t} = -\frac{\partial}{\partial \mathbf{r}} \cdot \mathbf{j}. \tag{10.23}$$

Integrating both sides over a spatial region R with an unmoving boundary ∂R, we obtain

$$\frac{\partial Q(R)}{\partial t} = \frac{\partial}{\partial t} \int_R \rho \, d^3 r = - \int_R \frac{\partial}{\partial \mathbf{r}} \cdot \mathbf{j} \, d^3 r = - \int_{\partial R} \mathbf{j} \cdot d\Sigma. \qquad (10.24)$$

On the far right we made use of Gauss's law (9.37). Equation (10.24) tells us that the rate at which the charge $Q(R)$ inside R increases equals minus the rate at which charge flows outward across the boundary ∂R of R. It is, in other words, a statement of the conservation of charge. Equation (10.23) is known as an *equation of continuity*.

10.4 A fuzzy metric

Our next task is to link the effects of gravity to their causes. Once again we *assume* that we are dealing with particles and nothing but particles and that, therefore, the behavior of particles (including their aggregates) is affected by nothing but particles (including their aggregates).

So far the components g_{ik} of the metric have been taken to be in possession of exact values. Yet, again, particles are *fuzzy*: neither their positions nor their momenta are in possession of exact values. If the metric represents effects that particles have on particles, its components cannot be sharp.

But now we run into a problem. As long as the metric is sharp, its components can be treated as functions of spacetime points, at least in our mathematical imagination. If the metric becomes fuzzy, so do the distances between spacetime points. But it is distances that are physically accessible, not spacetime points *per se*. Physically meaningful locations are *defined* by the distances between them. If spacetime distances become fuzzy, it becomes inconsistent to define the metric over a set of points that is locally isomorphic with a set of quadruplets of real numbers—points that can be labeled by spacetime coordinates. Physicists have not yet learned how to successfully circumnavigate this conundrum. Perhaps this is the reason—or at least one of the reasons—why we do not have a renormalizable quantum theory of gravity yet.

Since the present theory's lack of renormalizability does not prevent us from expressing the fuzziness of the metric in terms of path integrals, we can again use the principle of least action to obtain the classical theory. Assuming that the particles and the potentials follow fixed paths in their respective configuration spacetimes, we look for the conditions in which the action is stationary under arbitrary infinitesimal variations of the metric.

To be able to do so, we need a gravity term analogous to the term (10.7):

$$S_\mathcal{G} = \frac{1}{c} \int \mathcal{L}(g^{ik}, \partial g^{ik}/\partial x^l)\sqrt{-g}\, d^4x\,. \qquad (10.25)$$

In the presence of the metric, the volume element is given by $\sqrt{-g}\, d^4x$. g is the *determinant* of the metric, which is given by

$$g = \epsilon^{ijkl}\, g_{0i}\, g_{1j}\, g_{2k}\, g_{3l}\,, \qquad (10.26)$$

where

$$\epsilon^{ijkl} = \begin{cases} +1 & \text{if } ijkl \text{ is an } even\ permutation \text{ of } 0123 \\ -1 & \text{if } ijkl \text{ is an } odd\ permutation \text{ of } 0123 \\ 0 & \text{if } ijkl \text{ is not a } permutation \text{ of } 0123 \end{cases} \qquad (10.27)$$

As it turns out, there is no 4-scalar that can be constructed from the components of the metric and their first derivatives. But there is just one 4-scalar that additionally depends on the components' second derivatives. This is equal to the sum of two terms, one that depends on the components g_{ik} and their first derivatives only, and one that can be converted into an irrelevant boundary integral—irrelevant for reasons analogous to those given in the paragraph following Eq. (10.17). This is the *curvature scalar*

$$\mathcal{R} \stackrel{\text{Def}}{=} g^{jl} R_{jl} = R^l{}_l\,, \qquad (10.28)$$

which is a *contraction* of the *Ricci tensor*

$$R_{jl} \stackrel{\text{Def}}{=} g^{ik} R_{ijkl} = R^k{}_{jkl}\,, \qquad (10.29)$$

which in turn is a contraction of the *Riemann (curvature) tensor*

$$R^i{}_{jkl} \stackrel{\text{Def}}{=} \frac{\partial \Gamma^i_{jl}}{dx^k} - \frac{\partial \Gamma^i_{jk}}{dx^l} + \Gamma^i_{mk}\Gamma^m_{jl} - \Gamma^i_{ml}\Gamma^m_{jk}\,. \qquad (10.30)$$

In Gaussian units, the wanted gravity term is therefore

$$S_\mathcal{G} = \frac{c^3}{16\pi G} \int \mathcal{R}\sqrt{-g}\, d^4x\,, \qquad (10.31)$$

where $G = 6.674 \times 10^{-11}\,\mathrm{m^3\,kg^{-1}s^{-2}}$ is Newton's gravitational constant.

10.4.1 Meaning of the curvature tensor

Imagine a quadrilateral whose sides are the infinitesimal vectors du^i and dv^j. Take a vector A^k and parallel transport it about this quadrilateral—first along du^i, then along dv^j, then along $-du^k$, and finally along $-dv^l$. The resultant change in A^i is given by

$$\delta A^i = -R^i{}_{klm} A^k\, du^l\, dv^m\,. \qquad (10.32)$$

In other words, $R^i{}_{klm}$ is a machine with three inputs and one output. Insert the two sides of the quadrilateral into the last two input slots; insert a vector that is transported about the quadrilateral into the first input slot; get (minus) the resultant change of this vector.

10.4.2 *Cosmological constant*

In the weak-curvature limit, the action (10.31) yields Newton's law of gravity. It we do not regard this as necessary, the above constraints on the form of the gravity term allow for the addition of a constant to the curvature scalar $\mathcal{R}$. The resulting action is usually written in this form:

$$S_{\mathcal{G}} = \frac{c^3}{16\pi G} \int (\mathcal{R} - 2\Lambda)\sqrt{-g}\, d^4x\,. \tag{10.33}$$

Λ is the so-called *cosmological constant*. According to recent astronomical observations, it is small but not zero. Einstein originally introduced it to allow for the possibility of a *static* universe. After the discovery, by Edwin Hubble, that the universe was in fact *expanding*, Einstein dismissed it as "the greatest blunder" of his life. Today the cosmological constant is seen as the most likely reason why the expansion of the universe is *accelerating* rather than slowing down, as one would expect from the mutual gravitational attraction of the known forms of matter.

10.5 Einstein's equation

The total action is now given by the sum $S_{\mathcal{G}} + S_{\mathcal{M}}$, where

$$S_{\mathcal{M}} = \frac{1}{c} \int \mathcal{L}_{\mathcal{M}} \sqrt{-g}\, d^4x\,. \tag{10.34}$$

$\mathcal{L}_{\mathcal{M}}$ is a function of the matter variables and their first derivatives. ("Matter" now stands for *everything but* the metric.) The classical equations that state the metric's dependence on the distribution and motion of matter are obtained by varying the metric while the matter variables are held fixed. The result is the Einstein equation,

$$R_{ik} - \frac{1}{2}\mathcal{R}g_{ik} + \Lambda g_{ik} = \frac{8\pi G}{c^4}\, T_{ik}\,, \tag{10.35}$$

which is to Einstein's theory of gravity—the general theory of relativity—what Maxwell's equations are to classical electrodynamics. The tensor T_{ik} depends on the specific form of $\mathcal{L}_{\mathcal{M}}$.

10.5.1   *The energy–momentum tensor*

In much the same way as Eq. (10.19) implies the conservation law (10.22) (Problem 10.9), Eq. (10.35) implies the conservation law

$$T^{ik}{}_{;k} = 0. \tag{10.36}$$

The latter, however, only holds locally. If we chose a spacetime point and a sufficiently small neighborhood containing it, we can (within this neighborhood) use inertial coordinates, so that (within this neighborhood) Eq. (10.36) takes the form of an equation of continuity—or, rather, four of them, one for each index i:

$$\frac{\partial T^{ik}}{\partial x^k} = 0. \tag{10.37}$$

The conservation of the right-hand side of Eq. (10.19) and the conservation of the right-hand side of Eq. (10.35) both are implied by the respective left-hand sides of these equation. There is however another reason why the tensor T_{ik} on the right-hand side of Eq. (10.35) is (locally) conserved: the matter action (10.34) can be shown to be invariant under transformations of the metric that are locally equivalent to spacetime translations. This justifies the identification of T_{ik} in Eq. (10.35) with the *energy–momentum tensor* of matter.

10.6   Aharonov–Bohm effect

Let us return to the two-slit experiment we discussed in Secs. 5.2–5.3. The interference term contained the factor $\cos(k\Delta)$ (Eq. 5.3), where Δ stood for $\overline{DR} - \overline{DL}$, the difference between the distances of the detector from the slits. Because the electron source is equidistant from the slits, Δ is also the difference between the lengths of the paths $G \to L \to D$ and $G \to R \to D$. We will call these paths $\mathcal{L}$ and $\mathcal{R}$, respectively.

If the electron is subject to effects of the type that is represented by the vector potential $\mathbf{A}$, then (by Eq. 7.8) the phase associated with $\mathcal{L}$ contains the additional term

$$\frac{q}{c\hbar} \int_{\mathcal{L}} \mathbf{A} \cdot d\mathbf{r}, \tag{10.38}$$

and the phase associated with $\mathcal{R}$ contains the additional term

$$\frac{q}{c\hbar} \int_{\mathcal{R}} \mathbf{A} \cdot d\mathbf{r}. \tag{10.39}$$

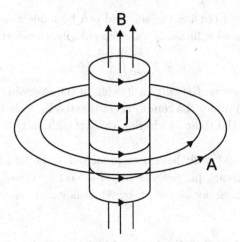

Fig. 10.1 The vector potential and the magnetic field associated with a current bearing solenoid.

The difference between these terms is an integral over the loop $G \to L \to D \to R \to G$,

$$\frac{q}{c\hbar} \int_{\mathcal{L}} \mathbf{A} \cdot d\mathbf{r} - \frac{q}{c\hbar} \int_{\mathcal{R}} \mathbf{A} \cdot d\mathbf{r} = \frac{q}{c\hbar} \oint \mathbf{A} \cdot d\mathbf{r} \stackrel{\text{Def}}{=} \phi. \qquad (10.40)$$

If a vector potential is present (in the equations), the interference term thus contains the factor $\cos(k\Delta + \phi)$. Invoking Stokes's theorem (Eq. 9.31) and the definition of the magnetic field (Eq. 9.48), we find that the integral of $\mathbf{A}$ along that loop equals the flux of $\mathbf{B}$ through any surface Σ bounded by it:

$$\oint \mathbf{A} \cdot d\mathbf{r} = \int_{\Sigma} \mathbf{B} \cdot d\Sigma. \qquad (10.41)$$

What happens if a long, thin solenoid passes through the loop $G \to L \to D \to R \to G$, at a right angle to the plane containing the loop, such that there is a clear distance between it and the loop? If a current is flowing in the solenoid, then there is—on paper— a magnetic field (Fig. 10.1), and if there is no electric field (or only a static one), then, according to Maxwell's equations (10.20), the curl of this magnetic field is related to the density of the current via

$$\frac{\partial}{\partial \mathbf{r}} \times \mathbf{B} = \frac{4\pi}{c} \mathbf{j}. \qquad (10.42)$$

The magnetic field outside the solenoid can be made arbitrarily weak by making the solenoid sufficiently long. Essentially, therefore,

$$\phi = \frac{q}{c\hbar} \int_\Sigma \mathbf{B} \cdot d\mathbf{\Sigma} . \qquad (10.43)$$

Because ϕ depends on the current flowing in the solenoid, the interference pattern observed under the conditions stipulated by Rule B can be shifted to the left or to the right by altering the strength and/or the direction of the current.

This effect could easily have been predicted in the late 1920's, by which time all the necessary physics was in place. Yet by many accounts it was predicted only three decades later, by Aharonov and Bohm (1959).[3]

10.7   Fact and fiction in the world of classical physics

Why did it take that long to predict such a remarkable effect? It is no excuse that classical electrodynamics can be formulated exclusively in terms of $\mathbf{E}$ and $\mathbf{B}$, nor that the four components of V and $\mathbf{A}$, which uniquely determine the six components of $\mathbf{E}$ and $\mathbf{B}$, are themselves not unique (Sec. 10.1). For it was well known that the Schrödinger equation could accommodate electromagnetic effects only in terms of the potentials V and $\mathbf{A}$ (see Eq. 7.23), and that loop integrals of $\mathbf{A}$ were gauge-invariant. Why was it nevertheless a widely held view that the potentials "have no physical meaning and are introduced solely for the purpose of mathematical simplification of the equations" [Rohrlich (1965)]?

What is implied by Rohrlich's remark is that $\mathbf{E}$ and $\mathbf{B}$ have a kind of physical meaning that the potentials lack. The general idea at the time was that the electromagnetic field is a physical entity in its own right; it is locally acted upon by charges, it locally acts on charges, and it mediates the action of charges on charges by locally acting on itself. At the heart of this notion is the so-called "principle of local action," felicitously articulated by DeWitt and Graham (1971) in an *American Journal of Physics* resource letter:

> Physicists are, at bottom, a naive breed, forever trying to come to terms with the "world out there" by methods which, however imaginative and refined, involve in essence the same element of contact as a well-placed kick.

---

[3]Actually it was first predicted *two* decades later [Ehrenberg and Siday (1949)], but it made a splash only after the publication of the paper by Aharonov and Bohm. This was followed by an actual demonstration of the effect in less than a year.

With the notable exception of Roger Boscovich, a Croatian physicist and philosopher who flourished in the 18th Century, it does not seem to have occurred to anyone that local action is as unintelligible as the apparent ability of material objects to act where they are not, which the principle of local action has purportedly explained away.[4]

If it is believed that electromagnetic effects on charges are locally produced by **E** and **B**, rather than at a distance by the distribution and motion of charges, then something like the Aharonov–Bohm effect cannot be foreseen, for neither **E** nor **B** is significantly different from zero along the two alternative electron paths. The lesson we have learned (or ought to have learned) from the history of this effect is that *neither* the potentials *nor* the fields should be thought of as physical entities in their own right. They both are computational tools.

What about gravitational effects? Here, too, the objects acted upon follow the geodesics of a differential geometry, and the consequences of being acted upon are represented by the curvature of this geometry. In the electromagnetic case, the curvature depends not only on the objects that act but also on the objects that are acted upon, through their masses and charges. In the gravitational case, the curvature is independent of the objects acted upon. Does this warrant the reification of the geometry representing gravitational effects? Or is it simply an instance of what mathematician and philosopher A.N. Whitehead (1997) has called "the fallacy of misplaced concreteness"?

Remember how we got to this point. We were looking for propagators for particles acted upon by particles. We found two mathematical constructs that could be used to incorporate into the propagators of particles the effects that particles have on particles—the 4-potential $(V, \mathbf{A})$ and the metric. To take into account the fuzziness of the particles acted upon, we summed over paths in their configuration spacetime. To take into account the fuzziness of the potentials and the metric, we summed over paths in *their* configuration spacetimes. We then obtained the laws of classical physics by taking the classical limit—the limit in which the fuzziness that "fluffs out" matter disappears and a single extremal path survives (for the particles as well as for the potentials and the metric). In this limit, the quantum-mechanical

---

[4]The impression that local action is intelligible derives from the familiarity of experiences like pulling a rope or pushing a stalled car. Why do the pushing hands not pass right through the car? Taking a microscopic look at what happens between the surface of the car and the surfaces of the pushing hands, we discover that this apparently local action involves net interatomic and intermolecular repulsive forces that act at a distance.

probability algorithms degenerate into trivial probability algorithms, which only assigns trivial probabilities (either 0 or 1).

A trivial probability algorithm—represented by a point in some phase space—can be interpreted as a state in the classical sense of the word: a collection of possessed properties (Sec. 8.1). What happens in the classical limit, then, is that the quantum laws, which correlate the probabilities of measurement outcomes *probabilistically*, degenerate into laws that correlate intrinsically possessed properties or values *deterministically*. And because deterministic correlations between intrinsically possessed properties or values lend themselves to causal interpretations, what happens is that the quantum-mechanical algorithms, which serve to compute probabilities of possible measurement outcomes on the basis of actual outcomes, degenerate into algorithms that serve to compute effects that matter has on matter. What they do not is degenerate into descriptions of physical mechanisms or processes by which matter acts on matter. As N. David Mermin (2009) reminisced near the end of a distinguished career:

> When I was an undergraduate learning classical electromagnetism, I was enchanted by the revelation that electromagnetic fields were real. Far from being a clever calculational device for how some charged particles push around other charged particles, they were just as real as the particles themselves, most dramatically in the form of electromagnetic waves, which have energy and momentum of their own and can propagate long after the source that gave rise to them has vanished.

> That lovely vision of the reality of the classical electromagnetic field ended when I learned as a graduate student that what Maxwell's equations actually describe are fields of operators on Hilbert space. Those operators are quantum fields, which most people agree are not real but merely spectacularly successful calculational devices. So real classical electromagnetic fields are nothing more (or less) than a simplification in a particular asymptotic regime (the classical limit) of a clever calculational device. In other words, classical electromagnetic fields are another clever calculational device.

10.7.1   *Retardation of effects and the invariant speed*

In Newton's theory, gravitational effects are instantaneous: the Earth is attracted towards the Sun's present position. It has been argued that this was the reason why Newton could not but refuse to "frame hypotheses" (about the mechanism by which gravity acts). Electromagnetic effects, on the other hand, are retarded. The earliest time at which a solar flare can affect us is about eight minutes later—the distance between the Sun and the Earth divided by the speed of light. According to a widely held belief, the retardation of electromagnetic effects is what made it possible to understand the mechanism by which these effects are transmitted.

Looking at it in another way, we reach the opposite conclusion. In Secs. 6.4–6.5 we established the necessity of an invariant speed. (Reminder: anything that "travels" with this speed in one inertial frame, will do so in every other inertial frame.) In Newton's theory the invariant speed is infinite; in a relativistic theory it is finite. But regardless, what matters is the existence of an invariant speed, inasmuch as it implies the existence of a special kind of spatiotemporal relation between events: simultaneity, which is absolute in Newton's case, or the lightlike relation, which is absolute in the relativistic case.

Suppose that an event e_1 at $(\mathbf{r}_1, t_1)$ is the cause of an event e_2 at $(\mathbf{r}_2, t_2)$. The fact that e_2 happens at t_2, rather than at any other time, has two possible explanations, depending on whether the causal connection between the events is mediated or unmediated. In the mediated case, t_2 is determined by the speed of mediation. This could be the speed of a material object traveling from $\mathbf{r}_1$ to $\mathbf{r}_2$, or the speed of a signal propagating in an elastic medium, or something more exotic. In the unmediated case, t_2 can only be determined by the special spatiotemporal relation that is implied by the existence of an invariant speed. In Newton's theory, t_2 is equal to t_1, while in a relativistic theory, t_2 is given by $t_1 + |\mathbf{r}_2 - \mathbf{r}_1|/c$. A delay of e_2 by $t_2 - t_1 = |\mathbf{r}_2 - \mathbf{r}_1|/c$ thus ought to be seen as the signature of an unmediated causal connection between e_1 and e_2, rather than as an indication that the causal relation between e_1 and e_2 is mediated by some physical process.

Again, in a non-relativistic theory, in which the stratification of space-time into hypersurfaces of constant time ("simultaneities") is absolute, energy and momentum are globally conserved: the total energy is the same in all simultaneities, as is the total momentum. In a relativistic theory, where the stratification of spacetime into simultaneities is frame-dependent, the law of conservation of energy–momentum holds for every stratification. Energy and momentum must therefore, in some sense, be locally conserved.[5]

---

[5] In a relativistic theory, energy–momentum cannot disappear in one place and instantly re-appear in a different place, inasmuch as this would only hold with respect to one frame or class of frames. Described using a different frame, it would re-appear not only in a different place but also at a different time, so that energy-momentum would not even be globally conserved.

But in what sense? Like every local conservation law, the conservation of energy–momentum is a feature of the mathematical tools we employ to calculate the correlations *between measurement outcomes*—both the probabilistic correlations of quantum physics and the deterministic correlations of classical physics. For instance, it ensures that in all particle "collision" experiments, and regardless of the reference frame used, the total energy–momentum of the incoming particles equals the total energy–momentum of the outgoing particles. (If some energy–momentum escapes undetected, then the following conditional is warranted: if the escaped energy–momentum were detected, it would agree with the local conservation law.)

But why give a thought to the meaning of the approximate laws of classical physics, considering that *"philosophically we are completely wrong with the approximate law,"* as Richard Feynman stressed at the beginning of his famous Caltech lectures [Feynman *et al.* (1963), original emphasis]? While learning the classical laws, most students of physics absorb a considerable amount of metaphysical embroidery that does not bear scrutiny in light of the underlying quantum laws. If this embroidery is not seen for what it is and discarded in time, it needlessly frustrates students' efforts to make sense of quantum physics. Mermin (2009) was able to rid himself of "our habit of inappropriately reifying our successful abstractions," but how many of us are?

Chapter 11

Quantum mechanics resumed

11.1 The experiment of Elitzur and Vaidman

In the following experiment [Elitzur and Vaidman (1993)] a Mach–Zehnder interferometer will be used. This consists of two beam splitters (BS_1 and BS_2), two mirrors (M_1 and M_2), and two photodetectors (D_1 and D_2) arranged as in Fig. 11.1. A particular twist of this experiment is the possible presence of a "bomb"—a photodetector that will explode if it absorbs a photon. For simplicity's sake we make the usual assumption that all detectors (including the bomb) are 100% efficient.

Imagine, to begin with, that neither BS_2 nor the bomb is present. A photon beam enters BS_1 from the left. Classically described, two beams emerge, each with half the intensity of the incoming beam. Described in quantum-mechanical terms, each photon has a 50% chance of being detected by D_1 (indicating that the photon was reflected upward by BS_1) and an equal chance of being detected by D_2 (indicating that the photon went horizontally through BS_1).

If BS_2 (but as yet no bomb) is present, Rule B applies. Here is what we need to know about the amplitudes associated with the alternatives (reflection by M_1 or reflection by M_2): they are equal *except* that each reflection causes a phase shift of $\pi/2$. (The magnitude of the phase shift depends on the materials used. For the sake of convenience, we use materials for which it equals $\pi/2$.) Such a phase shift is equivalent to the inclusion of a factor i.

Each of the alternatives leading to D_1 involves two reflections, so the corresponding amplitudes are equal ($i^2 A$, say). The probability of detection by D_1 thus is $p_1^B = |-A-A|^2 = 4|A|^2$. The alternative leading to D_2 via M_1 involves three reflections, so that the corresponding amplitude has an extra factor $i^3 = -i$. The alternative leading to D_2 via M_2 involves a single

139

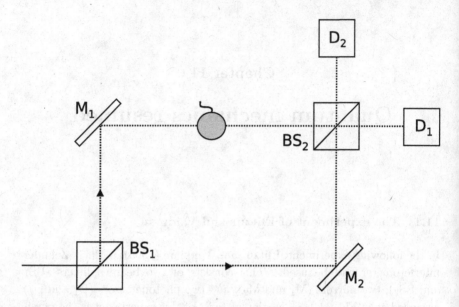

Fig. 11.1   "Bomb testing" experiment of Elitzur and Vaidman.

reflection, and the corresponding amplitude has an extra factor $+i$. Because the two amplitudes differ by a factor -1, the probability of detection by D_2, p_2^B, equals 0. (Under the conditions stipulated by Rule A, we would have that $p_1^A = p_2^A = 2|A|^2$.)

Finally, if both BS_2 and the bomb are present, the alternative taken by the photon is measured. If the bomb explodes, we can conclude that the photon went via M_1, and if it does not explode, we can conclude that the photon went via M_2. If it went via M_2, either photodetector responds with probability $1/2$. Thus:

- If the bomb is *absent*, D_1 clicks *every time* (100%), and D_2 *never* clicks (0%).
- If the bomb is *present*, it explodes half of the time (50%), and if it doesn't explode, D_1 and D_2 are *equally likely* to respond (25% each).

Problem 11.1. *Suppose that the bomb is present. Is it possible, with the help of the present setup, to ascertain the presence of the bomb without setting it off? Ponder this before you proceed.*

The answer is affirmative, albeit only in 25% of the tests. If the bomb explodes, which happens in 50% of the tests, we have failed. If the bomb is present and D_1 responds, which happens in 25% of the tests, we have learned nothing, for D_1 also responds if the bomb is absent. But if D_2 responds, which happens in 25% of the tests, we have succeeded, for D_2 never responds if the bomb is absent.

When a version of this experiment was demonstrated at a science fair in Groningen, the Netherlands, in 1995 [du Marchie van Voorthuysen (1996)], the reactions of non-physicists differed markedly from those of physicists. Everyone was perplexed, for the detection of the photon by D_2 seems to have contradictory implications:

- The bomb was present.
- The photon never came near the bomb.

If the photon never came near the bomb, how was it possible to learn that the bomb was present? While most ordinary folks thought that some physicist will eventually solve this puzzle, the physicists themselves were decidedly less hopeful that a satisfactory explanation will be found.

11.2 Observables

Many students of quantum mechanics are more or less apodictically informed that observables are self-adjoint operators, and that the possible values of an observable are its eigenvalues. These statements would be virtually self-evident if they were accompanied by sufficient emphasis on the probabilistic nature of the mathematical tools of quantum mechanics, which is rarely the case.

Consider a measurement whose possible outcomes are represented by projectors $|v_k\rangle\langle v_k|$, $k = 1, 2, \ldots$. If the density operator is $\hat{\mathbf{W}}$, the corresponding probabilities are $\langle v_k|\hat{\mathbf{W}}|v_k\rangle$, and their mean value is

$$\langle v \rangle = \sum_k v_k \langle v_k|\hat{\mathbf{W}}|v_k\rangle. \tag{11.1}$$

If the density operator is pure, $\hat{\mathbf{W}} = |w\rangle\langle w|$, the probabilities are $\langle v_k|w\rangle \langle w|v_k\rangle = \langle w|v_k\rangle \langle v_k|w\rangle$, and their mean value is

$$\langle v \rangle = \langle w|\hat{\mathbf{V}}|w\rangle \quad \text{with} \quad \hat{\mathbf{V}} = \sum_k |v_k\rangle v_k \langle v_k|. \tag{11.2}$$

Remembering the spectral theorem (8.30), we conclude that $\hat{\mathbf{V}}$ is self-adjoint, that the vectors $|v_k\rangle$ are its eigenvectors, and that the values v_k are the corresponding eigenvalues.

11.3   The continuous case

When dealing with continuous observables,[1] it is often more convenient to work with the *position representation* $\psi(x) = \langle x|\psi\rangle$ of a Dirac vector $|\psi\rangle$ than to work with the Dirac vector itself. For example, we may write the mean value (4.14) either as

$$\langle x\rangle = \langle\psi|\hat{\mathbf{X}}|\psi\rangle \quad \text{with} \quad \hat{\mathbf{X}} = \int |x\rangle\, x\, \langle x|\, dx$$

or as

$$\langle x\rangle = \int \psi^*(x)\,\hat{x}\,\psi(x)\, dx = \int \langle\psi|x\rangle\,\hat{x}\,\langle x|\psi\rangle\, dx \quad \text{with} \quad \hat{x} = x\,.$$

Using the position representation has the advantage that applying $\hat{x}$ to $\psi(x)$ boils down to multiplying $\psi(x)$ by x.

Another representation convenient to work with is the *momentum representation* $\psi(k) = \langle k|\psi\rangle$ of a Dirac vector $|\psi\rangle$. It allows us to cast the mean value (4.16) into the form

$$\langle p\rangle = \int \langle\psi|k\rangle\,\hat{p}\,\langle k|\psi\rangle\, dk \tag{11.3}$$

with $\hat{p} = \hbar k$. The momentum operator in the position representation has to satisfy

$$\langle p\rangle = \int \langle\psi|x\rangle\,\hat{p}\,\langle x|\psi\rangle\, dx\,.$$

We will now demonstrate with the help of Eq. (4.9) that $\hat{p} = (\hbar/i)(\partial/\partial x)$ fits the bill.

$$\langle p\rangle = \int \psi^*(x)\,\frac{\hbar}{i}\,\frac{\partial}{\partial x}\,\psi(x)\, dx$$

$$= \frac{1}{2\pi}\int \left[\int \overline{\psi}^*(k',t)\,e^{-ik'x}dk'\right]\frac{\hbar}{i}\,\frac{\partial}{\partial x}\left[\int \overline{\psi}(k,t)\,e^{ikx}dk\right]dx$$

$$= \frac{1}{2\pi}\int \left[\int \overline{\psi}^*(k',t)\,e^{-ik'x}dk'\right]\left[\int \overline{\psi}(k,t)\,\hbar k\,e^{ikx}dk\right]dx$$

$$= \int\int \overline{\psi}^*(k',t)\,\hbar k\,\overline{\psi}(k,t)\left[\frac{1}{2\pi}\int e^{i(k-k')x}dx\right]dk'\, dk\,.$$

Because the square bracket in the last line is a way of writing the delta distribution $\delta(k - k')$ (Sec. 8.12.2), we end up with

$$\langle p\rangle = \int\int \overline{\psi}^*(k',t)\,\hbar k\,\overline{\psi}(k,t)\,\delta(k - k')\, dk'\, dk = \int \overline{\psi}^*(k,t)\,\hbar k\,\overline{\psi}(k,t)\, dk\,.$$

This is Eq. (11.3) with $\hat{p} = \hbar k$.

---

[1] It would be more correct to speak of observables whose possible measurement outcomes form a denumerable set of mutually disjoint subsets of a non-denumerable set.

Problem 11.2. *The functions $\psi_k(x,t)$ (Eq. 4.7) satisfy the "continuum normalization"*

$$\int dx\, \psi_k^*(x,t)\, \psi_{k'}(x,t) = \delta(k - k').$$

11.4   Commutators

Let $\hat{\mathbf{U}}$ and $\hat{\mathbf{V}}$ be the self-adjoint operators associated with two discrete observables U and V. Introducing the abbreviations $\hat{\mathbf{U}}_i = |u_i\rangle\langle u_i|$ and $\hat{\mathbf{V}}_k = |v_k\rangle\langle v_k|$ for the projectors representing the possible outcomes of measurements of U and V, we have that

$$\hat{\mathbf{U}} = \sum_i u_i \hat{\mathbf{U}}_i, \qquad \hat{\mathbf{V}} = \sum_k v_k \hat{\mathbf{V}}_k.$$

Introducing the *commutator*

$$[\hat{\mathbf{A}}, \hat{\mathbf{B}}] \stackrel{\text{Def}}{=} \hat{\mathbf{A}}\hat{\mathbf{B}} - \hat{\mathbf{B}}\hat{\mathbf{A}} \tag{11.4}$$

and inserting the above expressions, we obtain

$$[\hat{\mathbf{U}}, \hat{\mathbf{V}}] = \sum_i \sum_k u_i v_k\, [\hat{\mathbf{U}}_i, \hat{\mathbf{V}}_k]. \tag{11.5}$$

According to Postulate 2 (Sec. 8.7), the outcomes of compatible elementary tests correspond to commuting projectors. Each of the projectors $\hat{\mathbf{U}}_i$ and $\hat{\mathbf{V}}_k$ represents the outcome of an elementary test. If these projectors commute, the operators $\hat{\mathbf{U}}$ and $\hat{\mathbf{V}}$ commute, and (measurements of) U and V are said to be compatible. If $\hat{\mathbf{U}}$ and $\hat{\mathbf{V}}$ fail to commute, (measurements of) U and V are said to be incompatible.

Problem 11.3. $(*)$ $[\hat{x}, \hat{p}] = i\hbar$.

The uncertainty relation (4.18) can also be deduced from this commutator [e.g., Marchildon (2002); McMahon (2006)].

Problem 11.4.

$$[\hat{\mathbf{A}}, \hat{\mathbf{B}}\hat{\mathbf{C}}] = [\hat{\mathbf{A}}, \hat{\mathbf{B}}]\, \hat{\mathbf{C}} + \hat{\mathbf{B}}\, [\hat{\mathbf{A}}, \hat{\mathbf{C}}].$$

11.5 The Heisenberg equation

Suppose $\hat{O}$ is the self-adjoint operator associated with an observable O. Let us differentiate $\langle\psi_2|\hat{O}|\psi_1\rangle$ with respect to time, making use of Eq. (8.48):

$$\frac{d\langle\psi_2|\hat{O}|\psi_1\rangle}{dt} = \left\langle -\frac{i}{\hbar}\hat{H}\psi_2\middle|\hat{O}\middle|\psi_1\right\rangle + \left\langle\psi_2\middle|\frac{\partial\hat{O}}{\partial t}\middle|\psi_1\right\rangle + \left\langle\psi_2\middle|\hat{O}\middle|-\frac{i}{\hbar}\hat{H}\psi_1\right\rangle$$

$$= \frac{i}{\hbar}\left\langle\psi_2\middle|\hat{H}\hat{O}\middle|\psi_1\right\rangle + \left\langle\psi_2\middle|\frac{\partial\hat{O}}{\partial t}\middle|\psi_1\right\rangle - \frac{i}{\hbar}\left\langle\psi_2\middle|\hat{O}\hat{H}\middle|\psi_1\right\rangle$$

$$= \left\langle\psi_2\middle|\frac{\partial\hat{O}}{\partial t} + \frac{i}{\hbar}[\hat{H},\hat{O}]\middle|\psi_1\right\rangle. \tag{11.6}$$

Since this holds for arbitrary vectors, we are entitled to look upon the operator sandwiched between the two vectors in the last line as the total time derivative of $\hat{O}$:

$$\frac{d\hat{O}}{dt} \stackrel{\text{Def}}{=} \frac{\partial\hat{O}}{\partial t} + \frac{i}{\hbar}[\hat{H},\hat{O}]. \tag{11.7}$$

This is the *Heisenberg equation.* We gather from it that O is a conserved quantity if the associated operator (i) has no explicit time dependence and (ii) commutes with the Hamiltonian $\hat{H}$.

11.6 Operators for energy and momentum

We have yet to show that

$$\hat{p} = \frac{\hbar}{i}\frac{\partial}{\partial\mathbf{r}} \tag{11.8}$$

is the momentum operator for the position representation.[2] We begin by observing that a continuous transformation $|\psi\rangle \to |\psi'\rangle$ that leaves Born probabilities unchanged,

$$|\langle\psi_2'|\psi_1'\rangle|^2 = |\langle\psi_2|\psi_1\rangle|^2, \tag{11.9}$$

is linear and unitary [cf. Sec. 8.12.1; Peres (1995) pp. 217–220; Wigner (1997) p. 233]. Since the infinitesimal translation $|\mathbf{r}\rangle \to |\mathbf{r}+\delta\mathbf{r}\rangle$ does not affect the probability

$$\left|\int\psi_2^*(\mathbf{r})\,\psi_1(\mathbf{r})\,d^3r\right|^2,$$

---

[2]In Sec. 11.3 we showed that $\hat{p} = (\hbar/i)(\partial/\partial x)$ is the momentum operator for the position representation *if* $\hat{p} = \hbar k$ is the same for the momentum representation. The following proof confirms that $\hat{p} = \hbar k$ is indeed the momentum operator for the momentum representation.

it is therefore effected by a unitary operator $\hat{U}(\delta\mathbf{r}) = 1 - i\hat{\mathbf{A}} \cdot \delta\mathbf{r}$:

$$\langle\psi|\mathbf{r} + \delta\mathbf{r}\rangle = \langle\psi|1 - i\hat{\mathbf{A}} \cdot \delta\mathbf{r}|\mathbf{r}\rangle = \langle\psi|\mathbf{r}\rangle - i\langle\psi|\hat{\mathbf{A}}|\mathbf{r}\rangle \cdot \delta\mathbf{r}. \qquad (11.10)$$

Since the components of $\hat{\mathbf{A}}$ are self-adjoint operators $\hat{\mathbf{A}}_x, \hat{\mathbf{A}}_y, \hat{\mathbf{A}}_z$, we can use $\langle\psi|\hat{\mathbf{A}}|\mathbf{r}\rangle^* = \langle\mathbf{r}|\hat{\mathbf{A}}|\psi\rangle$ to obtain the complex conjugate of Eq. (11.10):

$$\psi(\mathbf{r} + \delta\mathbf{r}) = \psi(\mathbf{r}) + i\langle\mathbf{r}|\hat{\mathbf{A}}|\psi\rangle \cdot \delta\mathbf{r}. \qquad (11.11)$$

We also have that

$$\psi(\mathbf{r} + \delta\mathbf{r}) = \psi(\mathbf{r}) + \frac{\partial\psi}{\partial\mathbf{r}} \cdot \delta\mathbf{r}, \qquad (11.12)$$

so that

$$\langle\mathbf{r}|\hat{\mathbf{A}}|\psi\rangle = \frac{1}{i}\frac{\partial\psi}{\partial\mathbf{r}}. \qquad (11.13)$$

Using the continuum version $\int |\mathbf{r}\rangle\langle\mathbf{r}| \, d^3r$ of the identity operator, we obtain the expected value

$$\langle\psi|\hat{\mathbf{A}}|\psi\rangle = \int \langle\psi|\mathbf{r}\rangle\langle\mathbf{r}|\hat{\mathbf{A}}|\psi\rangle \, d^3r = \int \psi^*(\mathbf{r})\frac{1}{i}\frac{\partial}{\partial\mathbf{r}} \psi(\mathbf{r}) \, d^3r. \qquad (11.14)$$

If the Hamiltonian $\hat{H}$ has no explicit dependence on $\mathbf{r}$—as is the case if we are dealing with a closed system and using inertial coordinates—then it commutes with the operator $(1/i)(\partial/\partial\mathbf{r})$, and the corresponding observable is conserved. In particular, if the Hamiltonian associated with a particle has no explicit dependence on $\mathbf{r}$, space is homogeneous as far as this particle is concerned. But the observable that is conserved on account of the homogeneity of space is *momentum*. The momentum operator for the position representation is thus given by Eq. (11.8). (To give it the conventional units of momentum, we included the factor $\hbar$.)

By the same route we arrive at the energy operator for the position representation:

$$\hat{\mathbf{E}} = i\hbar\frac{\partial}{\partial t}. \qquad (11.15)$$

The reason for the sign difference between the two operators is the same as that cited in Sec. 9.3.

11.7 Angular momentum

Using polar coordinates (Fig. 11.2), we can define the operator $(\hbar/i)(\partial/\partial\phi)$ in analogy with the operator $(\hbar/i)(\partial/\partial x)$. Because this operator has no explicit dependence on time, the corresponding observable is conserved if the

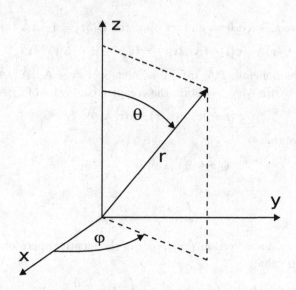

Fig. 11.2   Polar coordinates.

Hamiltonian has no explicit dependence on ϕ and therefore commutes with $(\hbar/i)(\partial/\partial\phi)$. In particular, if the Hamiltonian associated with a particle has no explicit dependence on ϕ, space is isotropic with respect to the z axis as far as this particle is concerned. But the observable that is conserved on account of the isotropy of space with respect to the z axis, is the z component of *angular momentum*.

To return to rectangular coordinates, we evaluate

$$\frac{\partial\psi}{\partial\phi} = \frac{\partial\psi}{\partial y}\frac{\partial y}{\partial\phi} + \frac{\partial\psi}{\partial x}\frac{\partial x}{\partial\phi}.$$

Since $x = r\sin\theta\cos\phi$ and $y = r\sin\theta\sin\phi$, as we gather from Fig. 11.2, this works out at $\partial\psi/\partial\phi = x(\partial\psi/\partial y) - y(\partial\psi/\partial x)$. Hence

$$(\hbar/i)(\partial/\partial\phi) = \hat{x}\,\hat{p}_y - \hat{y}\,\hat{p}_x \overset{\text{Def}}{=} \hat{L}_z. \tag{11.16}$$

The operators for the remaining angular momentum components are

$$\hat{L}_x = \hat{y}\,\hat{p}_z - \hat{z}\,\hat{p}_y, \qquad \hat{L}_y = \hat{z}\,\hat{p}_x - \hat{x}\,\hat{p}_z. \tag{11.17}$$

Problem 11.5. $(*)$

$$[\hat{L}_x, \hat{L}_y] = i\hbar\,\hat{L}_z, \quad [\hat{L}_y, \hat{L}_z] = i\hbar\,\hat{L}_x, \quad [\hat{L}_z, \hat{L}_x] = i\hbar\,\hat{L}_y. \tag{11.18}$$

Problem 11.6. $(*)$ *The operator* $\hat{\mathbf{L}}^2 \overset{\text{Def}}{=} \hat{L}_x^{\,2} + \hat{L}_y^{\,2} + \hat{L}_z^{\,2}$ *commutes with* $\hat{L}_x$, $\hat{L}_y$, *and* $\hat{L}_z$.

11.8 The hydrogen atom in brief

Using the time-independent Schrödinger equation (4.25) with $E_P = -e^2/r$, we find—as one would expect from the discussion in Sec. 4.4—that bound states exist only for specific values of E. These are exactly the values Bohr obtained in 1913 (Eq. 2.7):

$$E_n = -\frac{1}{n^2}\frac{m_e\,e^4}{2\hbar^2}, \qquad n = 1,2,3,\dots$$

The influence of the electron on the proton can be taken into account by substituting the *reduced mass* $\mu = m_p\,m_e/(m_p+m_e)$ for m_e, m_p being the proton's mass.

If polar coordinates are used, factorizing $\psi(r,\phi,\theta)$ into $e^{(i/\hbar)\,l_z\phi}\,\psi(r,\theta)$ leads to a ϕ-independent Schrödinger equation and a discrete set of values for l_z, just as factorizing $\psi(t,\mathbf{r})$ into $e^{-(i/\hbar)\,Et}\,\psi(\mathbf{r})$ led to a t-independent Schrödinger equation and a discrete set of values E_n. The ϕ-independent Schrödinger equation contains a real parameter whose possible values are given by $l(l+1)\hbar^2$, where l is an integer satisfying the condition $0 \le l \le n-1$. The possible values of l_z, in turn, are integers satisfying the inequality $|l_z| \le l$. The possible combinations of the *quantum numbers* n, l, l_z are thus

$$
\begin{array}{lll}
n=1 & l=0 & l_z=0 \\
n=2 & l=0 & l_z=0 \\
 & l=1 & l_z=0,\pm1 \\
n=3 & l=0 & l_z=0 \\
 & l=1 & l_z=0,\pm1 \\
 & l=2 & l_z=0,\pm1,\pm2 \\
\vdots & \vdots & \vdots
\end{array}
$$

The energy corresponding to the *principal quantum number* n is an eigenvalue of the Hamiltonian. The value $l(l+1)\hbar^2$, where l is the *angular momentum (or orbital, or azimuthal) quantum number*, is an eigenvalue of $\hat{\mathbf{L}}^2$. And the angular momentum component with respect to the z axis, $\hbar l_z$, is an eigenvalue of $\hat{\mathbf{L}}_z$. l_z is usually called the *magnetic quantum number*, hence the letter m is often used instead. The eigenfunctions $\psi_{nlm}(r,\phi,\theta)$ of these operators form a complete set of bound-state solutions $(E < 0)$ of the Schrödinger equation for the hydrogen atom.

As "a picture is worth more than a thousand words" (and as the relevant mathematics is set out in great detail in many textbooks), we content ourselves with the illustrations on the following pages. States with $l = 0,1,2,3$ were originally labeled s, p, d, f (for "sharp," "principal," "diffuse," and "fundamental") with a view to characterizing the corresponding spectral

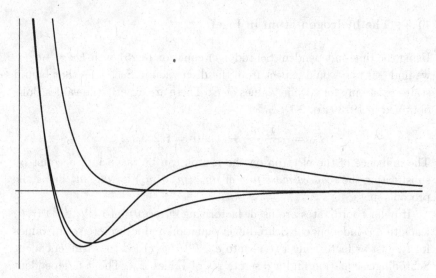

Fig. 11.3   Radial dependence of the first three spherically symmetric stationary states of atomic hydrogen. Their quantum numbers nlm are $1s0$, $2s0$, and $3s0$.

lines. States with higher l follow the alphabet (omitting the letters already used). Figure 11.3 maps the radial dependences of the first three spherically symmetric stationary states, which do not depend on ϕ or θ. Figures 11.4 and 11.5 plot the position probability distributions defined by some of the stationary states that are symmetric about the z axis. Figure 11.4 emphasizes the fuzziness of these *orbitals* at the expense of their rotational symmetry. By plotting surfaces of constant probability, Fig. 11.5 emphasizes their 3-dimensional shape at the expense of their fuzziness.

It should be stressed that what we see in these images is neither the nucleus nor the electron but the fuzzy position of the electron relative to the nucleus. Nor do we see this fuzzy position "as it is." What we see is the plot of a position probability distribution. This is defined by outcomes of measurements determining the values of n, l, and m, and it defines a fuzzy position by determining the probabilities of the possible outcomes of a subsequent position measurement. (Take any region R of the imaginary space of sharp positions relative to the proton, integrate this distribution over R, and obtain the probability of finding the electron in R.)

Since the dependence on ϕ is contained in a phase factor $e^{im\phi}$, it cannot be seen in plots of $|\psi(r,\phi,\theta)|^2$. To make this dependence visible, it is customary to replace $e^{im\phi}$ by its real or imaginary part, as has been done in Fig. 11.6.

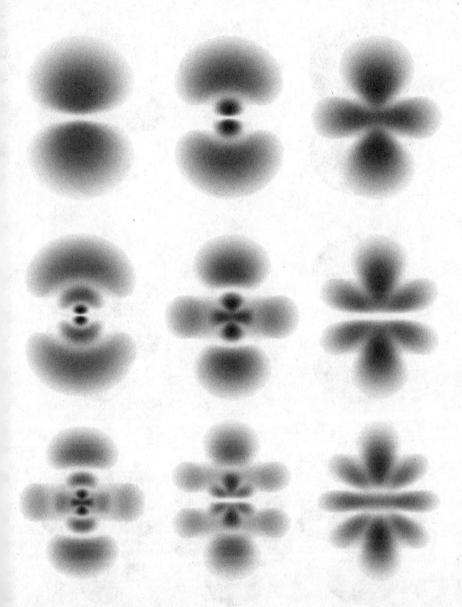

Fig. 11.4   The position probability distributions associated with the following orbitals. First row: 2p0, 3p0, 3d0. Second row: 4p0, 4d0, 4f0. Third row: 5d0, 5f0, 5g0. Images created with Orbital Viewer 1.04 by David Manthey.

Fig. 11.5   The same position probability distributions as in Fig. 11.4 differently rendered. Images created with Orbital Viewer 1.04 by David Manthey.

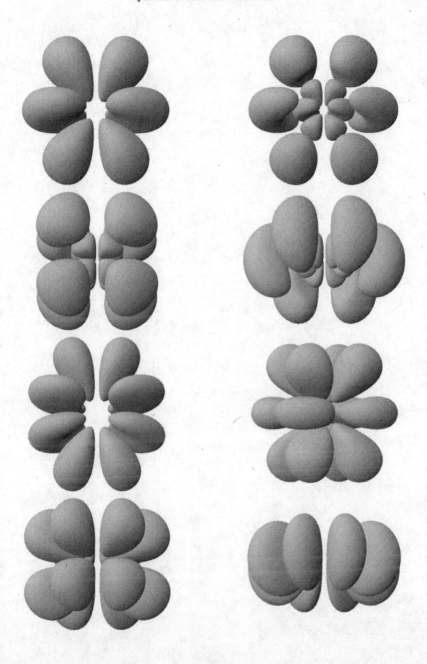

Fig. 11.6   Orbitals with $m \neq 0$. The squares of their *real parts* are shown. First row: $4f1$, $5f1$. Second row: $5f2$, $5f3$. Third row: $5g1$, $5g3$. Fourth row: $5g3$, $5g4$.

Chapter 12

Spin

12.1 Spin 1/2

In 1922, Otto Stern and Walther Gerlach found that when a narrow beam of silver atoms is sent through an inhomogeneous magnetic field, it splits into two beams. If the gradient of the magnetic field is oriented parallel to the z axis ("upward"), and if the incoming beam runs parallel to the y axis, then some atoms are deflected upward by a certain angle and the rest are deflected downward by the same angle. The measurement of the angle of deflection is repeatable: if the atoms in the upper (lower) beam are made to pass through a second, identical apparatus, all of them are deflected upward (downward).

Let the unit vectors $|z_+\rangle$ and $|z_-\rangle$ stand for the possible outcomes of this measurement, in lieu of the projectors $|z_+\rangle\langle z_+|$ and $|z_-\rangle\langle z_-|$. Representing different outcomes of the same measurement, the two vectors are orthogonal, and corresponding to a complete set of possible outcomes, they form a basis in a 2-dimensional vector space. For reasons that will soon become clear, we refer to the property whose value is indicated by this measurement as "the z component of the atom's spin."

The gradient of the magnetic field can of course point in any direction. If it is parallel to the x axis, we are set to measure the x component of the atom's spin. Its possible values are represented by the vectors $|x_+\rangle$ and $|x_-\rangle$. A different complete set of mutually orthogonal unit vectors, they form another basis in this 2-dimensional space of spin states. By Born's rule, the probability of obtaining the outcome $|z_+\rangle$ after having obtained the outcome $|x_+\rangle$ (and made sure that no external influence has affected the probabilities of the possible outcomes in the meantime) is $|\langle z_+|x_+\rangle|^2$.

153

According to a common phraseology, the first measurement *prepares* the atom in the state $|x_+\rangle$, with the result that subsequently the atom is *in* this state. Expressions of this sort are seriously misleading. For one thing, a quantum state is a probability algorithm, and a physical system cannot be "in" a probability algorithm.

Let us send the atoms in the upper beam through an apparatus that is identical to the first except that it is rotated by an angle α about the z axis. We shall denote by $|z_+''\rangle$ and $|z_-''\rangle$ the possible outcomes of a measurement made with this apparatus. Since the two apparatuses measure the same spin component, the probabilities $|\langle z_+''|z_+\rangle|^2$ and $|\langle z_-''|z_-\rangle|^2$ are equal to unity, and the probabilities $|\langle z_-''|z_+\rangle|^2$ and $|\langle z_+''|z_-\rangle|^2$ are zero. The amplitude $\langle z_+''|z_+\rangle$ is thereby determined up to a phase factor.

Problem 12.1. (∗) *If* $\langle z_+''|z_+\rangle = e^{i\phi}$ *then* $|z_+''\rangle = e^{-i\phi}|z_+\rangle$.

Our next move is to interpose a third measurement of the z component of the atom's spin, using an apparatus that is rotated relative to the first by an angle $\alpha/2$ about the z axis. The corresponding outcomes are $|z_+'\rangle$ and $|z_-'\rangle$. If this intermediate measurement is *not* made, then by Rule B we have that

$$\langle z_+''|z_+\rangle = \langle z_+''|z_+'\rangle\langle z_+'|z_+\rangle + \langle z_+''|z_-'\rangle\langle z_-'|z_+\rangle. \tag{12.1}$$

But we also have that $|\langle z_-'|z_+\rangle|^2 = |\langle z_+''|z_-'\rangle|^2 = 0$. This leaves us with $\langle z_+''|z_+\rangle = \langle z_+''|z_+'\rangle\langle z_+'|z_+\rangle$ or

$$e^{i\phi(\alpha)} = e^{i\phi(\alpha/2)}e^{i\phi(\alpha/2)}. \tag{12.2}$$

Since the right-hand side equals $e^{i\,2\phi(\alpha/2)}$, ϕ is proportional to α: $\langle z_+''|z_+\rangle = e^{ib\alpha}$. By the same token, $\langle z_-''|z_-\rangle = e^{ib'\alpha}$. Because overall phase factors lack physical significance—for at the end of the day we are left with absolute squares of sums of amplitudes—we can see to it that

$$|z_+''\rangle = e^{-il\alpha}|z_+\rangle, \qquad |z_-''\rangle = e^{il\alpha}|z_-\rangle. \tag{12.3}$$

What do we know about the value of l? For one thing, it cannot be zero. Every vector in the space of spin states is a linear combination of $|z_+\rangle$ and $|z_-\rangle$. If these two vectors are not affected by a rotation about the z axis—i.e., if $|z_+''\rangle = |z_+\rangle$ and $|z_-''\rangle = |z_-\rangle$—then no vector is affected, nor are the probabilities of outcomes of spin measurements with respect to the x and y axes changed. But these probabilities ought to be changed by a rotation about the z axis.

Nor can l be equal to unity, for if it were, a rotation by π about the z axis would change the signs of $|z_+\rangle$ and $|z_-\rangle$ and, consequently, those of all vectors. All probabilities would again remain unchanged as a result. Yet the smallest angle of rotation that leaves all probabilities unchanged ought to be 2π. This tells us that l equals $1/2$.

Problem 12.2. *With $l = 1/2$, a rotation by 2π about the z axis changes the sign of every vector.*

Particles with a 2-dimensional space of spin states are said to have a spin equal to $1/2$. This is the case if spin is measured in its natural units, which are dimensionless. Although spin has no classical analogue, it is related to angular momentum. Its conventional units are therefore those of angular momentum (which, as you will recall, are also the conventional units of action). In these units, the spin of a silver atom equals $\hbar/2$.

The reason why particles of spin $1/2$ are of particular interest to us is that they comprise all known constituents of "ordinary" matter—electrons and nucleons or electrons and quarks.

12.1.1 *Other bases*

To find out how the vectors $|x_\pm\rangle$ and $|y_\pm\rangle$ are related to the vectors $|z_\pm\rangle$, we begin by considering the effect on $|z_\pm\rangle$ of two successive rotations by π about the y axis. Each rotation inverts the z axis, so only a phase factor can appear in addition to the obvious inversions from $|z_\pm\rangle$ to $|z_\mp\rangle$ and back to $|z_\pm\rangle$:

$$|z_+\rangle \xrightarrow{\pi|y} e^{i\beta}|z_-\rangle \xrightarrow{\pi|y} e^{i\beta}e^{i\gamma}|z_+\rangle,$$

$$|z_-\rangle \xrightarrow{\pi|y} e^{i\gamma}|z_+\rangle \xrightarrow{\pi|y} e^{i\gamma}e^{i\beta}|z_-\rangle.$$

Because a rotation by 2π changes the sign of every vector, $e^{i(\beta+\gamma)}$ must be equal to -1. The convention is to set $\beta = 0$, so that $e^{i\beta} = 1$ and $e^{i\gamma} = -1$:

$$|z_+\rangle \xrightarrow{\pi|y} |z_-\rangle, \qquad |z_-\rangle \xrightarrow{\pi|y} -|z_+\rangle. \qquad (12.4)$$

Equipped with this information, we now consider two consecutive rotations by $\pi/2$ about the y axis:

$$|z_+\rangle \xrightarrow{\frac{\pi}{2}|y} c|z_+\rangle + d|z_-\rangle \xrightarrow{\frac{\pi}{2}|y} c^2|z_+\rangle + cd|z_-\rangle + de|z_+\rangle + df|z_-\rangle,$$

$$|z_-\rangle \xrightarrow{\frac{\pi}{2}|y} e|z_+\rangle + f|z_-\rangle \xrightarrow{\frac{\pi}{2}|y} ec|z_+\rangle + ed|z_-\rangle + fe|z_+\rangle + f^2|z_-\rangle.$$

To agree with Eq. (12.4), the coefficients must satisfy the following conditions:

$$c^2 + de = 0, \quad cd + df = 1, \quad ce + ef = -1, \quad de + f^2 = 0.$$

The first and the last condition imply that $f = c$ or $f = -c$. The second condition pins it down to $f = c$ and yields $d = 1/2c$, and the third condition yields $e = -1/2c$. With these results, the first condition tells us that $c^4 = 1/4$, which has four solutions. We are free to choose $c = 1/\sqrt{2}$, and this gives us

$$|z_+\rangle \xrightarrow{\frac{\pi}{2}|y} \frac{1}{\sqrt{2}}\Big[|z_+\rangle + |z_-\rangle\Big], \quad |z_-\rangle \xrightarrow{\frac{\pi}{2}|y} \frac{1}{\sqrt{2}}\Big[-|z_+\rangle + |z_-\rangle\Big]. \quad (12.5)$$

12.1.2 *Rotations as 2×2 matrices*

Our next task will be to cast these rotations into matrix notation. If $|1\rangle$ and $|2\rangle$ make up a basis, any vector can be written in the form

$$|a\rangle = a_1|1\rangle + a_2|2\rangle, \quad (12.6)$$

and any linear operator $\hat{\mathbf{A}}$ can be given the form

$$\hat{\mathbf{A}} = A_{11}|1\rangle\langle1| + A_{12}|1\rangle\langle2| + A_{21}|2\rangle\langle1| + A_{22}|2\rangle\langle2|. \quad (12.7)$$

Applying $\hat{\mathbf{A}}$ to $|a\rangle$ (from the left), or inserting $|a\rangle$ into $\hat{\mathbf{A}}$ (from the right), and making use of the orthonormality conditions (8.9), we obtain

$$|b\rangle \stackrel{\text{Def}}{=} \hat{\mathbf{A}}|a\rangle = (A_{11}a_1 + A_{12}a_2)|1\rangle + (A_{21}a_1 + A_{22}a_2)|2\rangle. \quad (12.8)$$

Here is the same equation in matrix notation:

$$\begin{pmatrix} b_1 \\ b_2 \end{pmatrix} = \begin{pmatrix} A_{11} & A_{12} \\ A_{21} & A_{22} \end{pmatrix} \begin{pmatrix} a_1 \\ a_2 \end{pmatrix} = \begin{pmatrix} A_{11}a_1 + A_{12}a_2 \\ A_{21}a_1 + A_{22}a_2 \end{pmatrix}. \quad (12.9)$$

It is just another way of writing $b_i = \sum_{k=1}^{2} A_{ik}a_k$ for $i = 1, 2$. (It would not have escaped your notice that the dependence of the components of vectors and matrices on a particular basis is no longer explicit in this notation.)

Problem 12.3. $A_{ik} = \langle i|\hat{\mathbf{A}}|k\rangle$.

Problem 12.4. (∗) $\langle i|\hat{\mathbf{A}}^\dagger|k\rangle = A_{ki}^*$.

Problem 12.5. (∗) *The first column of the matrix representation of $\hat{\mathbf{A}}$ contains the components of $\hat{\mathbf{A}}|1\rangle$, the second those of $\hat{\mathbf{A}}|2\rangle$.*

With respect to the $|z_\pm\rangle$ basis, a rotation by π about the y axis (Eq. 12.4) is effected by

$$\hat{\mathbf{R}}(\pi|y) = \begin{pmatrix} 0 & -1 \\ 1 & 0 \end{pmatrix}, \quad (12.10)$$

while a rotation by $\pi/2$ about the y axis (Eq. 12.5) is effected by

$$\hat{\mathbf{R}}\left(\tfrac{\pi}{2}\big|y\right) = \begin{pmatrix} 1/\sqrt{2} & -1/\sqrt{2} \\ 1/\sqrt{2} & 1/\sqrt{2} \end{pmatrix}. \tag{12.11}$$

Problem 12.6. (∗) *Cast the transformation (6.43) into matrix notation.*

The multiplication of two $N \times N$ matrices $\hat{\mathbf{A}}, \hat{\mathbf{B}}$ yields an $N \times N$ matrix $\hat{\mathbf{C}} = \hat{\mathbf{A}} \cdot \hat{\mathbf{B}}$ whose components are defined by $C_{ik} = \sum_{j=1}^{N} A_{ij} B_{jk}$. Note that matrices don't commute: in general, $\hat{\mathbf{A}} \cdot \hat{\mathbf{B}} \neq \hat{\mathbf{B}} \cdot \hat{\mathbf{A}}$.

Problem 12.7. (∗) *In terms of matrix components, Eq. (8.38) reads*

$$\sum_{j=1}^{N} U_{ji}^{*} U_{jk} = \delta_{ik}. \tag{12.12}$$

Problem 12.8. *The matrices (12.10) and (12.11) satisfy*

$$\hat{\mathbf{R}}\left(\tfrac{\pi}{2}\big|y\right) \cdot \hat{\mathbf{R}}\left(\tfrac{\pi}{2}\big|y\right) = \hat{\mathbf{R}}(\pi|y). \tag{12.13}$$

Here is the upshot of Sec. 12.1 in matrix notation:

$$\hat{\mathbf{R}}\left(\alpha|z\right) = \begin{pmatrix} e^{-i\alpha/2} & 0 \\ 0 & e^{i\alpha/2} \end{pmatrix}. \tag{12.14}$$

Problem 12.9. $\hat{\mathbf{R}}(\alpha|z) \cdot \hat{\mathbf{R}}(\alpha|z) = \hat{\mathbf{R}}(2\alpha|z)$.

The isotropy of space argues (i) that this holds for any axis, and (ii) that for any axis,

$$\hat{\mathbf{R}}(2\pi) = \begin{pmatrix} -1 & 0 \\ 0 & -1 \end{pmatrix}. \tag{12.15}$$

Problem 12.10. *The matrices (12.10) and (12.15) satisfy*

$$\hat{\mathbf{R}}(\pi|y) \cdot \hat{\mathbf{R}}(\pi|y) = \hat{\mathbf{R}}(2\pi). \tag{12.16}$$

Problem 12.11.

$$\hat{\mathbf{R}}\left(\tfrac{3}{2}\pi\big|y\right) = \begin{pmatrix} -1/\sqrt{2} & -1/\sqrt{2} \\ 1/\sqrt{2} & -1/\sqrt{2} \end{pmatrix}. \tag{12.17}$$

One does not have to look far to find the right continuous functions of α that reproduce the components of the matrices (12.10), (12.11), (12.15), and (12.17):

$$\hat{\mathbf{R}}\left(\alpha|y\right) = \begin{pmatrix} \cos(\alpha/2) & -\sin(\alpha/2) \\ \sin(\alpha/2) & \cos(\alpha/2) \end{pmatrix}. \tag{12.18}$$

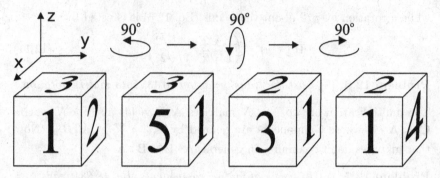

Fig. 12.1 A rotation by 90° about the x axis obtained by three consecutive rotations about the y and z axes.

Let us use this result to calculate the probability of finding the spin of a spin-1/2 particle "up" with respect to one axis after having found it "up" with respect to another axis (this can only depend on the angle between the two axes):

$$|\langle z_+|\hat{\mathbf{R}}\,(\alpha|y)\,|z_+\rangle|^2 = \cos^2(\alpha/2)\,. \tag{12.19}$$

As one would expect, this probability decreases continuously from 1 (for $\alpha = 0$) to 0 (for $\alpha = \pi$).

To find out how to rotate spin vectors about the x axis, we begin by performing the following sequence of rotations: first a rotation by $\pi/2$ about the z axis, next a rotation by $\pi/2$ about the y axis, and last a rotation by $-\pi/2$ about the z axis:

$$\begin{pmatrix} e^{i\pi/4} & 0 \\ 0 & e^{-i\pi/4} \end{pmatrix} \begin{pmatrix} \frac{1}{\sqrt{2}} & -\frac{1}{\sqrt{2}} \\ \frac{1}{\sqrt{2}} & \frac{1}{\sqrt{2}} \end{pmatrix} \begin{pmatrix} e^{-i\pi/4} & 0 \\ 0 & e^{i\pi/4} \end{pmatrix} = \begin{pmatrix} \frac{1}{\sqrt{2}} & -\frac{i}{\sqrt{2}} \\ -\frac{i}{\sqrt{2}} & \frac{1}{\sqrt{2}} \end{pmatrix}\,. \tag{12.20}$$

As we gather from Fig. 12.1, the result is a rotation by $\pi/2$ about the x axis.

Problem 12.12. *Verify that the product of the three matrices works out as stated.*

One also doesn't have to look far to find the right continuous functions of α that reproduce the components of the following matrices:

$$\hat{\mathbf{R}}\left(\tfrac{\pi}{2}|x\right) \begin{pmatrix} \frac{1}{\sqrt{2}} & -\frac{i}{\sqrt{2}} \\ -\frac{i}{\sqrt{2}} & \frac{1}{\sqrt{2}} \end{pmatrix}, \quad \hat{\mathbf{R}}\left(\pi|x\right) \begin{pmatrix} 0 & -i \\ -i & 0 \end{pmatrix},$$

$$\hat{\mathbf{R}}\left(\tfrac{3}{2}\pi|x\right) \begin{pmatrix} -\frac{1}{\sqrt{2}} & -\frac{i}{\sqrt{2}} \\ -\frac{i}{\sqrt{2}} & -\frac{1}{\sqrt{2}} \end{pmatrix}, \quad \hat{\mathbf{R}}\left(2\pi|x\right) \begin{pmatrix} -1 & 0 \\ 0 & -1 \end{pmatrix}.$$

Here they are:

$$\hat{\mathbf{R}}\left(\alpha|x\right) = \begin{pmatrix} \cos(\alpha/2) & -i\sin(\alpha/2) \\ -i\sin(\alpha/2) & \cos(\alpha/2) \end{pmatrix}. \tag{12.21}$$

We are now in a position to define the basis vectors with respect to the x and y axes by rotating $|z_+\rangle$ by the appropriate angle $(+\pi/2$ or $-\pi/2)$ about the appropriate axis:

$$|x_+\rangle = \hat{\mathbf{R}}\left(\tfrac{\pi}{2}|y\right)|z_+\rangle = \frac{1}{\sqrt{2}}\Big[|z_+\rangle + |z_-\rangle\Big], \tag{12.22}$$

$$|y_+\rangle = \hat{\mathbf{R}}\left(-\tfrac{\pi}{2}|x\right)|z_+\rangle = \frac{1}{\sqrt{2}}\Big[|z_+\rangle + i|z_-\rangle\Big], \tag{12.23}$$

$$|x_-\rangle = \hat{\mathbf{R}}\left(-\tfrac{\pi}{2}|y\right)|z_+\rangle = \frac{1}{\sqrt{2}}\Big[|z_+\rangle - |z_-\rangle\Big], \tag{12.24}$$

$$|y_-\rangle = \hat{\mathbf{R}}\left(\tfrac{\pi}{2}|x\right)|z_+\rangle = \frac{1}{\sqrt{2}}\Big[|z_+\rangle - i|z_-\rangle\Big]. \tag{12.25}$$

12.1.3 *Pauli spin matrices*

Which self-adjoint operators correspond to the three spin components of a spin-1/2 particle? The eigenvectors of these operators are $|x_\pm\rangle$, $|y_\pm\rangle$, and $|z_\pm\rangle$, respectively, while the corresponding eigenvalues are $\pm1/2$ (in natural units) or $\pm\hbar/2$ (in conventional units). If we take the eigenvalues to be ±1 (for "up" and "down"), the wanted operators—in matrix notation with respect to the basis $|z_\pm\rangle$—are the *Pauli matrices*:

$$\hat{\sigma}_x = \begin{pmatrix} 0 & 1 \\ 1 & 0 \end{pmatrix}, \quad \hat{\sigma}_y = \begin{pmatrix} 0 & -i \\ i & 0 \end{pmatrix}, \quad \hat{\sigma}_z = \begin{pmatrix} 1 & 0 \\ 0 & -1 \end{pmatrix}. \tag{12.26}$$

Problem 12.13.

$$\hat{\sigma}_x|x_\pm\rangle = \pm|x_\pm\rangle, \quad \hat{\sigma}_y|y_\pm\rangle = \pm|y_\pm\rangle, \quad \hat{\sigma}_z|z_\pm\rangle = \pm|z_\pm\rangle. \tag{12.27}$$

Problem 12.14.

$$\hat{\sigma}_x\hat{\sigma}_x = \hat{\sigma}_y\hat{\sigma}_y = \hat{\sigma}_z\hat{\sigma}_z = -i\hat{\sigma}_x\hat{\sigma}_y\hat{\sigma}_z = \hat{1} \stackrel{\text{Def}}{=} \begin{pmatrix} 1 & 0 \\ 0 & 1 \end{pmatrix}. \tag{12.28}$$

Problem 12.15.

$$\hat{\sigma}_x\hat{\sigma}_y = i\hat{\sigma}_z, \quad \hat{\sigma}_y\hat{\sigma}_z = i\hat{\sigma}_x, \quad \hat{\sigma}_z\hat{\sigma}_x = i\hat{\sigma}_y. \tag{12.29}$$

Problem 12.16. (∗)

$$[\hat{\sigma}_x, \hat{\sigma}_y] = 2i\hat{\sigma}_z, \quad [\hat{\sigma}_y, \hat{\sigma}_z] = 2i\hat{\sigma}_x, \quad [\hat{\sigma}_z, \hat{\sigma}_x] = 2i\hat{\sigma}_y. \tag{12.30}$$

When we introduced the commutator of two operators (Sec. 11.4), we concluded that measurements of the corresponding observables are compatible if and only if the operators commute. Since the operators corresponding to different components of the same spin do not commute, different components of the same spin are incompatible. There is no state that assigns probability 1 to possible outcomes of measurements of more than one spin component.

For later use we record the general form of a rotation by α about an axis defined by the unit vector $\hat{\mathbf{n}} = (n_x, n_y, n_z)$ [Marchildon (2002), Sec. 4.6]:

$$\hat{\mathbf{R}}(\alpha|\hat{\mathbf{n}}) = \begin{pmatrix} \cos(\alpha/2) - in_z \sin(\alpha/2) & -(in_x + n_y)\sin(\alpha/2) \\ (-in_x + n_y)\sin(\alpha/2) & \cos(\alpha/2) + in_z \sin(\alpha/2) \end{pmatrix}$$

$$= \exp\left(-i\frac{\alpha}{2}\,\hat{\mathbf{n}}\cdot\hat{\boldsymbol{\sigma}}\right).$$

The components of $\hat{\boldsymbol{\sigma}}$ are the Pauli matrices (12.26), and the exponential is defined by its Taylor expansion. A further streamlining of the notation is achieved by introducing the vector $\boldsymbol{\alpha} = \alpha\hat{\mathbf{n}}$ and by defining $\mathbf{s} = \hat{\boldsymbol{\sigma}}/2$. This allows us to write

$$\hat{\mathbf{R}}(\boldsymbol{\alpha}) = e^{-i\,\boldsymbol{\alpha}\cdot\mathbf{s}}. \tag{12.31}$$

In terms of the components of $\mathbf{s}$, the commutators (12.30) take the form

$$[s^a, s^b] = i\epsilon^{abc}s^c, \tag{12.32}$$

where

$$\epsilon^{abc} = \begin{cases} +1 & \text{if } abc \text{ is an even permutation of } xyz \\ -1 & \text{if } abc \text{ is an odd permutation of } xyz \\ 0 & \text{if } abc \text{ is not a permutation of } xyz \end{cases} \tag{12.33}$$

12.2   A Stern–Gerlach relay

A beam of spin-1/2 atoms passes through three inhomogeneous magnetic fields (from the right to the left as in Fig. 12.2). The gradients in the first and the third magnet point upward (toward the tip of the upper pole, where the field is strongest), while the gradient in the second magnet points to the right (relative to the atom's direction of motion). Unless the atoms are specially prepared, the beam gets split three times: first into two beams that are deflected upward and downward, respectively, then into four beams two of which are deflected to the left and two to the right, and finally into eight beams, four of which are deflected upward and four downward. We thus expect the atoms to hit the screen in eight spots.

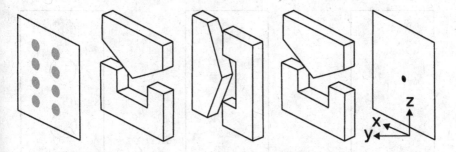

Fig. 12.2   The apparatus used in Sec. 12.2.

What happens if we gradually reduce the power of the second magnet, so that the four left spots first overlap and then merge with the four right spots? How many spots will be seen?

Let us calculate, assuming that initially the pure state $|a\rangle$ is associated with each atom. The amplitude for an atom's arrival in the uppermost right spot (with respect to the atom's direction of motion) is

$$\langle z_+|x_+\rangle\langle x_+|z_+\rangle\langle z_+|a\rangle = \frac{1}{2}\langle z_+|a\rangle. \qquad (12.34)$$

Problem 12.17. (∗) *The probability of an atom's arrival at the screen (no matter at which spot) equals 1.*

If the second magnet is switched off, the amplitude for an atom's arrival at the (now single) uppermost spot is

$$\langle z_+|x_+\rangle\langle x_+|z_+\rangle\langle z_+|a\rangle + \langle z_+|x_-\rangle\langle x_-|z_+\rangle\langle z_+|a\rangle = \langle z_+|a\rangle, \qquad (12.35)$$

whereas the amplitude for an atom's arrival at the (now single) spot directly below is

$$\langle z_-|x_+\rangle\langle x_+|z_+\rangle\langle z_+|a\rangle + \langle z_-|x_-\rangle\langle x_-|z_+\rangle\langle z_+|a\rangle = 0. \qquad (12.36)$$

By the same token, the amplitude for an atom's arrival at the (now single) lowermost spot is $\langle z_-|a\rangle$, whereas the amplitude for an atom's arrival at the (now single) spot directly above is 0. The reason why the atoms arrive in two spots rather than four is this: without the intermediate measurement of the x component, the repeatability of measurements guarantees that the outcome of the second measurement of the z component confirms the outcome of the first measurement. Figure 12.3 illustrates what happens if the power of the second magnet is gradually reduced.

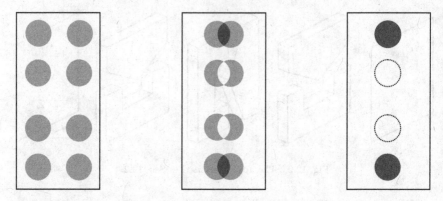

Fig. 12.3   Constructive or destructive interference occurs in regions where the outcome of the measurement of the spin's x component is no longer indicated.

12.3   Why spin?

This makes two questions: Why is it called "spin"? And what is it for?

In answer to the first question, we observe that wave functions transforming under rotations about the z axis as $\psi \longrightarrow e^{im\phi}\psi$ are eigenstates of $\hat{\mathbf{L}}_z$ with eigenvalue m, while spin states transforming under rotations about this axis as $|s\rangle \longrightarrow e^{i\phi/2}|s\rangle$ are eigenstates of $\hat{\sigma}_z$ with eigenvalue $1/2$. Both angular momentum and spin—also called *extrinsic* and *intrinsic* angular momentum, respectively—are conserved for closed systems because the Hamiltonians of such systems reflect the isotropy of space (in this particular case, with respect to rotations about the z axis).

And what is it for? The answer to this question emerges from one of those theorems that are as easy to state as they are hard to prove [Pauli (1940); Duck and Sudarshan (1998)]. Particle beams passing through an inhomogeneous magnetic field may split into $b = 2, 3, 4, \ldots$ beams depending on the particles' spin. Beams consisting of spinless particles, for which $l = 0$, don't split. Beams consisting of spin-$1/2$ particles split into two beams, as we have seen. The general rule is $b = 2\,l + 1$, where the possible values of l are $0, 1, 2, 3 \ldots$ and $1/2, 3/2, 5/2 \ldots$.

As we shall see in Sec. 14.2, every particle is either a *boson* or a *fermion*. The aforementioned theorem, due to Pauli, states that a fermion cannot have an integral spin ($l = 0, 1, 2, \ldots$), whereas a boson cannot have a half-integral spin ($l = 1/2, 3/2, 5/2, \ldots$). But only fermions obey the *exclusion principle* (Sec. 14.5), and the stability of matter requires that its fundamental constituents—electrons and nucleons or electrons and quarks—obey this principle (see Sec. 22.1). The long and the short of it is that the fundamental constituents of matter must be particles with a half-integral spin of at least 1/2.

12.4 Beyond hydrogen

The helium atom has two electrons, and its nucleus—two protons and two neutrons—has about four times the mass of the hydrogen nucleus. If the influence of the electrons on the nucleus as well as relativistic and spin effects are ignored, the corresponding stationary states are in principle obtainable as solutions of the following equation:

$$E \frac{\partial \psi}{\partial t} = - \frac{\hbar^2}{2m} \left[\frac{\partial^2 \psi}{\partial x_1^2} + \frac{\partial^2 \psi}{\partial y_1^2} + \frac{\partial^2 \psi}{\partial z_1^2} + \frac{\partial^2 \psi}{\partial x_2^2} + \frac{\partial^2 \psi}{\partial y_2^2} + \frac{\partial^2 \psi}{\partial z_2^2} \right]$$
$$+ \left[-\frac{2e^2}{r_1} - \frac{2e^2}{r_2} + \frac{e^2}{r_{12}} \right] \psi. \tag{12.37}$$

The wave function now depends on the six coordinates $x_1, y_1, z_1, x_2, y_2, z_2$ of the two electrons relative to the nucleus, and the potential energy is made up of three terms: two that are inverse proportional to the respective distances $r_1 = \sqrt{x_1^2 + y_1^2 + z_1^2}$ and $r_2 = \sqrt{x_2^2 + y_2^2 + z_2^2}$ between the electrons and the nucleus, and one that is inverse proportional to the distance $r_{12} = \sqrt{(x_2-x_1)^2 + (y_2-y_1)^2 + (z_2-z_1)^2}$ between the electrons.

The fact that we have one wave function depending on six coordinates rather than two separate wave functions each depending on three coordinates, should not come as a surprise. The sole purpose of wave functions depending on more than one coordinate is to let us calculate *joint* probabilities, which may be *correlated* (Sec. 1.4). The probability of finding the first electron in a region A and the second electron in a region B, for example, is given by the joint probability

$$p(A, B) = \int_A d^3 r_1 \int_B d^3 r_2 \, |\psi(\mathbf{r}_1, \mathbf{r}_2)|^2. \tag{12.38}$$

If the whereabouts of the two electrons were independent of each other, it would be possible to factorize $\psi(\mathbf{r}_1, \mathbf{r}_2)$ into $\psi_1(\mathbf{r}_1)$ and $\psi_2(\mathbf{r}_2)$, and $p(A, B)$ would be the product of the probabilities

$$p(A) = \int_A d^3 r_1 \, |\psi(\mathbf{r}_1)|^2, \quad p(B) = \int_B d^3 r_2 \, |\psi(\mathbf{r}_2)|^2.$$

But in general, and particularly where the solutions of Eq. (12.37) are concerned, this is not the case.

For the lowest energy levels, Eq. (12.37) has been solved by numerical methods. With three or more electrons in the picture, looking for exact solutions of the corresponding Schrödinger equation is a hopeless undertaking. Through further simplifications it is nevertheless possible to get valuable results. The Periodic Table and many properties of the chemical elements can be understood by using the *central field approximation* [e.g., Marchildon (2002), Sec. 9.3]. In this approach, one disregards the details of the interactions between electrons. One instead considers each electron subject to two potentials, one representing the effect of the nucleus, and one representing the effect of a continuous, spherically symmetric charge distribution, which does duty for the remaining electrons. Spin effects are also neglected *except* for the Pauli exclusion principle (Sec. 14.5), which implies that two electrons associated with the same wave function cannot be also associated with the same spin state. Measurements of their spin components with respect to the same axis—no matter which—will yield different outcomes. Because only two outcomes are possible, at most two electrons can share the same orbital.

The central field approximation yields stationary wave functions $\psi_{nlm}(\mathbf{r})$ for single electrons that are quite similar to those of atomic hydrogen, except that their dependence on the radial coordinate is modified by the charge distribution that stands in for the other electrons. This modification has the result that the energies associated with orbitals with the same quantum number n but with different quantum numbers l are no longer equal. For any $n > 1$, the mean distance between the electron and the nucleus increases with l. With a greater mean distance, the electron is more shielded against the nucleus by the cloud of negative charge representing the other electrons. Electrons with higher l are therefore less strongly bound and their *ionization energies* are lower.

Problem 12.18. *The number of orbitals with the same quantum number n is n^2.*

Chemists group orbitals into *electron shells*. Each shell encompasses all orbitals with the same principal quantum number n. The nth shell can therefore "accommodate" $2 \times n^2$ electrons—*twice* n^2 because each orbital can "accommodate" two electrons.

Helium has a "full" first shell and an "empty" second shell. Because the helium nucleus has twice the charge of the hydrogen nucleus, the two electrons are, on average, significantly closer to the nucleus than the single electron of the hydrogen atom. The ionization energy of helium, accordingly, is much higher, 24.6 eV as compared to 13.6 eV. On the other hand, if we tried to add an electron to create a negative helium ion, this would have to "go into" the second shell, which is strongly shielded from the nucleus by the electrons of the first shell. Helium is therefore neither prone to give up an electron not able to hold on to an extra electron. It is chemically inert, as are all of the elements in the rightmost column of the (traditional) Periodic Table.

Through the second row of the Periodic Table, the second shell gets "filled." Because the energies of the 2p orbitals are higher than the energy of the 2s orbital, the latter gets "occupied" first. With each added electron (and proton) the electron distribution as a whole gets pulled inward, and the ionization energy goes up, from 5.4 eV for lithium (atomic number $Z=3$) to 21.6 eV for neon ($Z=10$). Whereas lithium readily parts with its loosely bound outer electron, fluorine ($Z=9$), having a single "vacancy" in the second shell, is eager to grab one. Both are therefore quite active chemically. The progression from sodium ($Z=11$) to argon ($Z=18$) parallels that from lithium to neon.

There is a noteworthy peculiarity in this progression of ionization energies: the ionization energy of oxygen ($Z=8$, 13.6 eV) is lower than that of nitrogen ($Z=7$, 14.5 eV), and that of sulfur ($Z=16$, 10.36 eV) is lower than that of phosphorus ($Z=15$, 10.49 eV). To understand why this is so, we must take account of certain details that have so far been ignored. Suppose that one of the two 2p electrons of carbon ($Z=6$) goes into the $m=0$ orbital (with respect to the z axis). Where will the other 2p electron go? It will go into any "vacant" orbital that minimizes the repulsion between the two electrons by maximizing their mean distance. This is neither of the two orbitals with $|m|=1$ with respect to the z axis but an orbital with $m=0$ with respect to some axis perpendicular to the z axis. If we call this the x axis, then the third 2p electron of nitrogen goes into the orbital with $m=0$ relative to y axis. The fourth 2p electron of oxygen then has no choice but to go into an already occupied 2p orbital. This raises its energy sufficiently for the drop in ionization energy from nitrogen to oxygen.

By the time the 3p orbitals are completely "filled," the energies of the 3d states are pushed up so high (as a result of shielding) that the 4s state is energetically lower. The "filling" up of the 3d orbitals therefore begins only after the 4s orbitals are "full," with scandium ($Z=21$).

12.5 Spin precession

The Pauli equation (15.6) is to a non-relativistic particle of spin 1/2 what the Schrödinger equation (7.23) is to a non-relativistic particle without spin. The Pauli wave function has two components, and what interests us here is how they depend on each other:

$$i\hbar\frac{d}{dt}\begin{pmatrix}\psi_+ \\ \psi_-\end{pmatrix} = -\mu\,\hat{\boldsymbol{\sigma}}\cdot\mathbf{B}\begin{pmatrix}\psi_+ \\ \psi_-\end{pmatrix} = \begin{pmatrix} -\mu B_z & -\mu(B_x - iB_y) \\ -\mu(B_x + iB_y) & \mu B_z \end{pmatrix}\begin{pmatrix}\psi_+ \\ \psi_-\end{pmatrix}.$$

$\mathbf{B}$ is the magnetic field, and if q and m are the charge and mass of an electron, then $\mu = q\hbar/2mc$ is the *Bohr magneton*. If the particle is subject to a homogeneous magnetic field $\mathbf{B} = (0,0,B)$, the two equations are actually independent of each other:

$$i\hbar\frac{d\psi_+}{dt} = -\mu B\psi_+ \,, \quad i\hbar\frac{d\psi_-}{dt} = +\mu B\psi_- \,.$$

Their respective solutions are proportional to $e^{\pm(i/\hbar)\mu Bt}$, and they predict the respective outcomes $\pm\mu B$ for a measurement of the energy associated with the electron's spin. If instead $\mathbf{B} = (B,0,0)$, we obtain the coupled equations

$$i\hbar\frac{d\psi_+}{dt} = -\mu B\psi_- \,, \quad i\hbar\frac{d\psi_-}{dt} = -\mu B\psi_+ \,.$$

A possible pair of solutions is

$$\psi_+ = \cos\left(\frac{\mu B}{\hbar}t\right), \quad \psi_- = i\sin\left(\frac{\mu B}{\hbar}t\right).$$

This tells us that the probabilities associated with a measurement of the z component of the spin, p_+^z and p_-^z, oscillate as follows:

$$p_+^z = \cos^2\left(\frac{\mu B}{\hbar}t\right), \quad p_-^z = \sin^2\left(\frac{\mu B}{\hbar}t\right) = \cos^2\left(\frac{\mu B}{\hbar}t \pm \frac{\pi}{2}\right). \quad (12.39)$$

To find out how the probabilities p_+^y and p_-^y associated with a measurement of the y component of the spin oscillate, we make use of Eqs. (12.23) and (12.25):

$$\psi_+^y = \frac{1}{\sqrt{2}}\left(\psi_+ + i\psi_-\right)$$

$$= \frac{1}{\sqrt{2}}\left[\cos\left(\frac{\mu B}{\hbar}t\right) - \sin\left(\frac{\mu B}{\hbar}t\right)\right] = \cos\left(\frac{\mu B}{\hbar}t + \frac{\pi}{4}\right),$$

$$\psi_-^y = \frac{1}{\sqrt{2}}\left(\psi_+ - i\psi_-\right)$$

$$= \frac{1}{\sqrt{2}}\left[\cos\left(\frac{\mu B}{\hbar}t\right) + \sin\left(\frac{\mu B}{\hbar}t\right)\right] = \cos\left(\frac{\mu B}{\hbar}t - \frac{\pi}{4}\right).$$

Hence,

$$p_+^y = \cos^2\left(\frac{\mu B}{\hbar}t + \frac{\pi}{4}\right), \quad p_-^y = \cos^2\left(\frac{\mu B}{\hbar}t - \frac{\pi}{4}\right). \tag{12.40}$$

From Eqs. (12.39) and (12.40) we gather that the axis with respect to which a spin measurement yields "up" with probability 1 is precessing at the frequency $\omega = 2\mu B/\hbar$. (Bear in mind that the period of $\cos^2$ is π.)

12.6 The quantum Zeno effect

Consider a spin that is precessing about the x axis as described in the previous section. At $t = 0$, the probability of finding it up with respect to the z axis equals 1. At later times, this probability is given by $p(t) = \cos^2(\mu Bt/\hbar)$. If we measure the z component of the spin N times at intervals of duration t/N, the probability of finding it up *each time* equals

$$[p(t/N)]^N = \left[\cos^2(\mu Bt/N\hbar)\right]^N. \tag{12.41}$$

As $N \to \infty$, this tends to unity. A finite number of (instantaneous) measurements slows down the precession, and an infinite number of measurements performed during a finite interval would bring it to a halt.

While no measurement is instantaneous and the limit $N \to \infty$ is physically unattainable, the tendency of repeated measurements to slow down the rate at which the probabilities of the possible outcomes change, is a general and experimentally well-established feature of quantum mechanics [Misra and Sudarshan (1977); Peres (1980); Singh and Whitaker (1982)]. The effect is named after the Eleatic philosopher Zeno who, in the 5th Century B.C.E., put forth a series of apparent paradoxes designed to demonstrate that motion was impossible.

The reverse effect also exists. If there is no magnetic field and the spin is measured at intervals of duration t/N with respect to an axis that rotates with angular frequency ω about the x axis, the probability of finding it up *each time* is again given by Eq. (12.41). In this case, a finite number of (instantaneous) measurements causes a certain amount of precession, and in the unphysical limit $N \to \infty$ this approaches the angular frequency of the rotating apparatus. This too is a general and experimentally well-established feature of quantum mechanics. If an infinite number of measurements could be performed during a finite interval, the behavior of a quantum system would be determined exclusively by what the experimenters chose to measure.

The bottom line: for the quantum-mechanical correlation laws to be effective, not only must there be measurements (whose possible outcomes they serve to correlate) but also there must be unmeasured intervals between the measurements.

Chapter 13

Composite systems

13.1 Bell's theorem: The simplest version

Consider the following setup [Mermin (1985)] (Fig. 13.1):

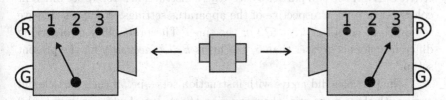

Fig. 13.1   Mermin's simplest version of Bell's theorem.

The device at the center launches two particles in opposite directions. Each particle enters an apparatus capable of performing one of three measurements. Each measurement has two possible outcomes, indicated by a red or green light. In each run of the experiment the measurement performed is randomly selected for each apparatus. After a large number of runs, we have in our hands a long record of apparatus settings and responses. This record has the following characteristics:

- Whenever both apparatuses perform the same measurement (11, 22, or 33), equal colors (RR or GG) are *never* observed.
- The pattern of R's and G's is *completely random*.

In particular, the apparatuses flash different colors exactly half of the time. If this does not bother you, then try to explain how it is that the colors differ whenever identical measurements are performed!

The obvious explanation is that each particle arrives with an "instruction set"—a set of properties determining how the apparatus will respond.

There are $2^3 = 8$ such sets: RRR, RRG, RGR, GRR, RGG, GRG, GGR, and GGG. If, for instance, a particle arrives with RGG, the apparatus flashes red if it is set to 1, and green if it is set to 2 or 3. According to this explanation, the reason why the outcomes differ whenever both particles are subjected to the same measurement is that the particles are launched with opposite instruction sets. If one particle carries the instruction set RRG, then the other particle carries the instruction set GGR.

Let us see if this explanation is tenable. Suppose that the instruction sets are RRG and GGR. In this case we expect to see different colors with five of the $3^2 = 9$ possible combinations of apparatus settings (namely, 11, 22, 33, 12, 21) and to see equal colors with four of them (namely, 13, 23, 31, and 32). Because the apparatus settings are randomly chosen, this pair of instruction sets produces different colors 5/9 of the time. The same is obviously true for the remaining pairs of instruction sets *except* the pair RRR, GGG. If the two particles carry these instruction sets, we see different colors *every time*, irrespective of the apparatus settings. If follows that we see different colors *at least* 5/9 of the time. The probability of observing different colors is greater than 5/9. This is *Bell's inequality* for the present setup.

If the particles did arrive with instruction sets—i.e., if each particle did come with three properties the possession of any one of which is *revealed* by the apparatus—then Bell's inequality would be satisfied. But it isn't, since the apparatuses flash different colors *half of the time*. We appear forced to conclude, in this instance as in many similar experimental situations, that the predictions of quantum mechanics cannot be explained with the help of instruction sets [Aspect (2002); Greenstein and Zajonc (1997); Laloë (2001); Redhead (1987)]. These measurements do not reveal *pre-existent* properties or values. In a radical sense, they *create* their outcomes. They create the properties or values whose possession (by a system or an observable) they indicate.

But then how is it that the colors differ whenever identical measurements are made? Since each apparatus indicates each possible outcome half of the time, the marginal probabilities of the outcomes are $p(R) = p(G) = 1/2$. If the outcomes were uncorrelated, we would have that $p(R,G) = p(G,R) = p(R,R) = p(G,G) = 1/4$. Instead we have $p(R,R) = p(G,G) = 0$ and $p(R,G) = p(G,R) = 1/2$ whenever the apparatus settings agree. What mechanism or process is responsible for these correlations? How does one apparatus or particle "know" which measurement is performed and which outcome is obtained by the other apparatus? I can tell you the answer straight off: you understand this as well as anybody else! As a distinguished Princeton physicist commented, "anybody who's not bothered by Bell's theorem has to have rocks in his head" [quoted by Mermin (1985)].

Einstein was bothered, albeit not by Bell's theorem, whose original version [Bell (1964)] appeared several years after his death. The title of Bell's 1964 paper, "On the Einstein Podolsky Rosen paradox," refers to a seminal paper of 1935, in which Einstein, Podolsky, and Rosen made use of similar correlations—now often collectively referred to as "EPR correlations"—to argue that quantum mechanics was incomplete. In 1947, Einstein wrote in a letter to Max Born that he could not seriously believe in the quantum theory "because it cannot be reconciled with the idea that physics should represent a reality in time and space, free from spooky actions at a distance" [Einstein (1971)]. In his 1964 paper, Bell was led to conclude that, on the contrary, "there must be a mechanism whereby the setting of one measurement device can influence the reading of another instrument, however remote." Spooky actions at a distance are here to stay! As Bell wrote in a subsequent paper [Bell (1966)], "the Einstein–Podolsky–Rosen paradox is resolved in a way which Einstein would have liked least."

13.2 "Entangled" spins

Consider two physical systems A and B, which have been subjected to measurements. The respective outcomes are $|a\rangle\langle a|$ and $|b\rangle\langle b|$. If no interactions take place between the systems—which could lead to correlations between the outcomes of subsequent measurements—then the systems remain associated with vectors $|a(t)\rangle$ and $|b(t)\rangle$ belonging to the respective vector spaces $\mathcal{V}_a$ and $\mathcal{V}_b$ of the individual systems. In this case the composite system made up of A and B is associated with the pair of vectors $|a(t)\rangle$ and $|b(t)\rangle$, which is itself a vector. This vector—denoted by $|a(t)\rangle \otimes |b(t)\rangle$ or, more simply, by $|a, b\,(t)\rangle$—belongs to the vector space $\mathcal{V}_a \otimes \mathcal{V}_b$, the *direct product* of the two vector spaces.

The scalar product in $\mathcal{V}_a \otimes \mathcal{V}_b$ is the product of two scalar products, one in $\mathcal{V}_a$ and one in $\mathcal{V}_b$:

$$\langle a', b'|a, b\rangle = \langle a'|a\rangle \, \langle b'|b\rangle. \tag{13.1}$$

If the vectors $|a_1\rangle, \ldots, |a_m\rangle$ form a basis in $\mathcal{V}_a$ and the vectors $|b_1\rangle, \ldots, |b_n\rangle$ form a basis in $\mathcal{V}_b$, then the mn vectors $|a_i, b_k\rangle$ form a basis in $\mathcal{V}_a \otimes \mathcal{V}_b$.

Given a composite system associated with the vector $|a, b\rangle$, the probability of obtaining the respective outcomes $|a'\rangle\langle a'|$ and $|b'\rangle\langle b'|$ thus equals

$$|\langle a', b'|a, b\rangle|^2 = |\langle a'|a\rangle|^2 \, |\langle b'|b\rangle|^2. \tag{13.2}$$

This is what we expect on the basis of Born's rule, and what motivates the above definitions in the first place.

Most vectors in $\mathcal{V}_a \otimes \mathcal{V}_b$ cannot be written as a single pair of vectors, and this is where it gets interesting. In such cases the outcomes of at least some of the measurements that can be performed on the component systems are correlated, and the component systems are said to be *entangled*. It is, however, always possible to find a pair of bases—say, $|A_i\rangle$ in $\mathcal{V}_a$ and $|B_j\rangle$ in $\mathcal{V}_b$—such that the vector associated with two entangled systems can be written as a *single* sum of *bi-orthogonal* terms [Peres (1995), Sec. 5.3]:

$$\sum_k c_k |A_k, B_k\rangle. \tag{13.3}$$

13.2.1 The singlet state

An example of such a vector is the *singlet state* of two spin-1/2 systems:

$$\| 0 \rangle\!\rangle = \frac{1}{\sqrt{2}} \Big(|z_+, z_-\rangle - |z_-, z_+\rangle \Big). \tag{13.4}$$

The double delimiters remind us that this vector belongs to $\mathcal{V}_a \otimes \mathcal{V}_b$.

Problem 13.1.

$$\big|\langle z_+, z_+ \| 0 \rangle\!\rangle\big|^2 = \big|\langle z_-, z_- \| 0 \rangle\!\rangle\big|^2 = 0, \tag{13.5}$$

$$\big|\langle z_+, z_- \| 0 \rangle\!\rangle\big|^2 = \big|\langle z_-, z_+ \| 0 \rangle\!\rangle\big|^2 = \frac{1}{2}. \tag{13.6}$$

Problem 13.2. *Using the rotation matrices of Sec. 12.1.2, show that* $\| 0 \rangle\!\rangle$ *is invariant under rotations about the x, y, and z axes.*

Problem 13.3. *Using Eqs. (12.22–12.25), show that*

$$\| 0 \rangle\!\rangle = \frac{-1}{\sqrt{2}} \Big(|x_+, x_-\rangle - |x_-, x_+\rangle \Big) = \frac{i}{\sqrt{2}} \Big(|y_+, y_-\rangle - |y_-, y_+\rangle \Big).$$

The singlet state $\| 0 \rangle\!\rangle$ is a two-particle state that can be used to reproduce the statistical properties described in Sec. 13.1. Here is how: The device at the center of Fig. 13.1 launches pairs of spin-1/2 particles in the singlet state. (Such a state can be obtained by letting a spinless particle decay into two spin-1/2 particles—for example, a π^0 meson into an electron and a positron. It can also be prepared by inducing a spinless hydrogen molecule to dissociate into a pair of hydrogen atoms or by letting two protons scatter each other at low energies.) The three observables that the apparatuses are designed to measure are the spin components with respect to three coplanar axes, each differing from the others by an angle of $2\pi/3$ (120°).

It is immediately clear that whenever the spins are measured with respect to the same axis (identical apparatus settings), opposite outcomes (different colors) are obtained. What remains to be shown is that the outcomes are completely random if the apparatus settings are not taken into account. As you will remember (Eq. 12.19), if the angle between two axes A_1 and A_2 is α, the probability of finding the spin of a spin-1/2 particle up with respect to A_2, after having found it up with respect to A_1, is given by $\cos^2(\alpha/2)$. Because of the negative correlations of the singlet state, this is also the probability of obtaining *opposite* outcomes if the spins of two particles in the singlet state are measured with respect to axes that differ by α.

Because the apparatus settings are randomly selected, the probability with which the same spin component is measured is $1/3$, and in these cases the probability of obtaining opposite results is 1. The probability with which the particle spins are measured with respect to different axes is $2/3$, and in these cases the probability of obtaining opposite results is $\cos^2(\pi/3) = 1/4$. If the apparatus setting are not taken into account, the probability of obtaining opposite outcomes is therefore $(1/3) \times 1 + (2/3) \times (1/4) = 1/2$.

13.3 Reduced density operator

An operator $\hat{\mathbf{A}} \otimes \hat{\mathbf{B}}$ acting on a vector $|a\rangle \otimes |b\rangle$ in $\mathcal{V}_a \otimes \mathcal{V}_b$ produces the vector $\hat{\mathbf{A}}|a\rangle \otimes \hat{\mathbf{B}}|b\rangle$. If two systems are associated with the bi-orthogonal decomposition (13.3), then according to Eq. (8.31) the joint probability of outcomes represented by the projectors $\hat{\mathbf{P}}_a$ (acting in $\mathcal{V}_a$) and $\hat{\mathbf{P}}_b$ (acting in $\mathcal{V}_b$) is

$$p(\hat{\mathbf{P}}_a, \hat{\mathbf{P}}_b) = \sum_i \sum_k c_i^* c_k \langle A_i, B_i | \hat{\mathbf{P}}_a \otimes \hat{\mathbf{P}}_b | A_k, B_k \rangle . \qquad (13.7)$$

If we set $\hat{\mathbf{P}}_b$ equal to the identity operator $\hat{\mathbf{1}}$ on $\mathcal{V}_b$, which represents the trivial "outcome" that provides no information whatever, we obtain the marginal probability

$$p(\hat{\mathbf{P}}_a) = \sum_i \sum_k c_i^* c_k \langle A_i, B_i | \hat{\mathbf{P}}_a \otimes \hat{\mathbf{1}} | A_k, B_k \rangle = \sum_i |c_i|^2 \langle A_i | \hat{\mathbf{P}}_a | A_i \rangle .$$

This can be written as

$$p(\hat{\mathbf{P}}_a) = \mathrm{Tr}(\hat{\mathbf{W}}_a \hat{\mathbf{P}}_a) \quad \text{with} \quad \hat{\mathbf{W}}_a = \sum_i |c_i|^2 |A_i\rangle\langle A_i| .$$

Equation (13.7), on the other hand, can be cast into the form

$$p(\hat{\mathbf{P}}_a, \hat{\mathbf{P}}_b) = \mathrm{Tr}\big(\hat{\mathbf{W}}(\hat{\mathbf{P}}_a \otimes \hat{\mathbf{P}}_b)\big) \quad \text{with} \quad \hat{\mathbf{W}} = \sum_i \sum_k c_i^* c_k |A_k, B_k\rangle\langle A_i, B_i| .$$

The *reduced density operator* $\hat{\mathbf{W}}_a$ for the first system can thus be obtained by taking a *partial trace* of the density operator for the composite system:

$$\hat{\mathbf{W}}_a = \sum_j \langle B_j | \left[\sum_i \sum_k c_i^* c_k |A_k, B_k\rangle\langle A_i, B_i| \right] |B_j\rangle . \tag{13.8}$$

Problem 13.4. *If two spin-1/2 particles are in the singlet state, the spin state of each individual particle is*

$$\hat{\mathbf{W}}_a = \frac{1}{2}\Big(|z_+\rangle\langle z_+| + |z_-\rangle\langle z_-|\Big) = \frac{1}{2}\,\hat{\mathbf{1}} . \tag{13.9}$$

We gather from this density operator that the possible outcomes of any spin measurement, with regard to any axis, are equally likely. The entire information provided by the singlet state therefore concerns correlations between outcomes of measurements performed on *both* particles.

13.4 Contextuality

In 1967, Simon Kochen and Ernst Specker proved a theorem that places constraints on the permissible types of hidden variables. Hidden variable theories posit an unobservable determinism that underlies the observed randomness of quantum physics. In the extreme case, a cryptodeterministic theory claims not only that all observables have definite values at all times but also that their values are independent of the devices by which they are measured. This idea goes back to the paper by Einstein *et al.* (1935), in which they considered the following criterion "reasonable":

> If, without in any way disturbing a system, we can predict with certainty (i.e., with probability equal to unity) the value of a physical quantity, then there exists an element of physical reality corresponding to this physical quantity.

In his 1951 textbook (Secs. 22.15–22.18), David Bohm mirrored the thought experiment discussed by EPR by looking at the dissociation of a spinless diatomic molecule. Since the dissociated atoms are in the singlet state, EPR could have argued that any spin component of the first atom can be measured by measuring the corresponding spin component of the second atom. After a sufficient amount of time, the two atoms may be assumed to be separated by a great distance, so EPR could have argued further that this measurement can be made without in any way disturbing the first atom. But if it is possible to determine the value of *any* spin component of an atom without disturbing it, then *all* of the atom's spin components must be in possession of definite values.

In Bohm's context, the preposterousness of EPR's claim jumps out at us right away. The possible values of any spin-component of a spin-1/2 particle are $+1/2$ and $-1/2$. If a positive value is predicted for the z component, then a negative value is predicted for the component with respect to the inverted z axis. If we continuously rotate the magnetic gradient of the apparatus from being parallel to the z axis to pointing in the opposite direction, the component with respect to the axis defined by the apparatus must at some point jump discontinuously from $+1/2$ to $-1/2$. Gleason (1957) and Bell (1966) formally disproved the possibility of such discontinuous jumps. Bell however went further, pointing out that Gleason's argument as well as his own "tacitly assumed that measurement of an observable must yield the same value independently of what other measurements may be made simultaneously." In other words, it assumed that measurements are *noncontextual*.

Kochen and Specker assumed that the pre-existent values of observables A, B, C posited by hidden variable theories mirror the algebraic relations between the corresponding self-adjoint operators. Specifically, the values $v(A), v(B), v(C)$ of A, B, C conform to the following constraints: if A, B, C are compatible, so that $\hat{A}, \hat{B}, \hat{C}$ commute, then

(1) $\hat{C} = \hat{A} + \hat{B}$ implies $v(C) = v(A) + v(B)$,
(2) $\hat{C} = \hat{A}\hat{B}$ implies $v(C) = v(A)\,v(B)$.

Kochen and Specker (1967) then showed, for vector spaces of at least three dimensions, that this seemingly innocuous assumption leads to the conclusion that hidden variables must in general be *contextual*. Their original proof, which is notoriously complex, used a set of 117 observables. Later it was shown that the smallest number of observables sufficient to establish the contextuality of hidden variables for 3-dimensional (and hence larger) vector spaces is 18 [Pavičić *et al.* (2005)]. Here I shall present a simpler proof, which uses the 4-dimensional state space associated with a system composed of two particles of spin 1/2 [Mermin (1990)].

Consider the following array of operators:

$$\begin{array}{ccc} \hat{1} \otimes \hat{\sigma}_z & \hat{\sigma}_z \otimes \hat{1} & \hat{\sigma}_z \otimes \hat{\sigma}_z \\ \hat{\sigma}_x \otimes \hat{1} & \hat{1} \otimes \hat{\sigma}_x & \hat{\sigma}_x \otimes \hat{\sigma}_x \\ \hat{\sigma}_x \otimes \hat{\sigma}_z & \hat{\sigma}_z \otimes \hat{\sigma}_x & \hat{\sigma}_y \otimes \hat{\sigma}_y \end{array} \tag{13.10}$$

Problem 13.5. *Using the three equations (12.27), show that each of these nine operators has the eigenvalues ±1.*

This means that the possible values of the corresponding observables are ±1.

Problem 13.6. (∗) *The three operators in each row and in each column commute.*

It follows that they can be measured simultaneously.

Problem 13.7. (∗) *In each row and each column, each operator is the product of the two others, except for the third column, where each operator is minus the product of the two others.*

Now assume that the observables corresponding to these operators have pre-existent values. Let us call them

$$\begin{array}{ccc} a_1 & a_2 & a_3 \\ b_1 & b_2 & b_3 \\ c_1 & c_2 & c_3 \end{array} \tag{13.11}$$

Since the operators in each row and each column of array (13.10) commute, we expect the corresponding values in each row and each column of array (13.11), all of which are equal to either +1 or −1, to satisfy the same multiplication rules as the corresponding operators. In other words, we expect that the following equations hold:

$$\begin{array}{ll} a_1 \, a_2 \, a_3 = 1 \,, & a_1 \, b_1 \, c_1 = 1 \,, \\ b_1 \, b_2 \, b_3 = 1 \,, & a_2 \, b_2 \, c_2 = 1 \,, \\ c_1 \, c_2 \, c_3 = 1 \,, & a_3 \, b_3 \, c_3 = -1 \,. \end{array} \tag{13.12}$$

But this is impossible: whereas the product of the left-hand sides of these six equations equals +1 (because each value occurs squared), the product of the right-hand sides equals −1.

The long and the short of it is that what we call "the outcome of a measurement of an observable A represented by the operator $\hat{A}$" cannot in general only depend on $\hat{A}$ and on the system on which the measurement is performed. It will also depend on the other measurements that are performed together with the measurement of A. It will be contextual.

Suppose, for example, that eight of the nine values in array (13.11) are as follows:

$$
\begin{array}{lll}
a_1 = 1 & a_2 = 1 & a_3 = 1 \\
b_1 = 1 & & b_3 = -1 \\
c_1 = 1 & c_2 = 1 & c_3 = 1
\end{array}
\qquad (13.13)
$$

If the observable having the pre-existent value b_2 is measured together with the observables having the pre-existent values $b_1 = 1$ and $b_3 = -1$, then $b_2 = -1$. If the same observable is measured together with the observables having the pre-existent values $a_2 = 1$ and $c_2 = 1$, then $b_2 = +1$. In Sec. 16.3 we shall return to the subject of contextuality for a final consideration.

13.5 The experiment of Greenberger, Horne, and Zeilinger

13.5.1 *A game*

In this section we start with a game. It involves two teams, the "players" (Andy, Bob, and Charles) and the "interrogators" [Vaidman (1999)]. According to the rules of this game,

(1) either all players are asked for the value of X,
(2) or one player is asked for the value of X, and the two other players are asked for the value of Y.

The possible values of both X and Y are $+1$ and -1. In case (1), the players win if and only if the product of their answers equals -1. In case (2), they win if and only if the product of their answers equals $+1$. Once the questions are asked, the players cannot communicate with each other. Before that, they may work out a strategy. Is there a fail-safe strategy? Can they make sure that they will win? Ponder this before you proceed.

The obvious strategy is to use pre-agreed answers. Let us call them X_A, X_B, X_C, Y_A, Y_B, Y_C.

Problem 13.8. *Assign values ± 1 to the following variables in such a way that the product of the three X values equals -1 and the product of every* *pair of Y values is equal to the X value of the remaining column—or else explain why this cannot be done.*

$$\begin{matrix} X_A & X_B & X_C \\ Y_A & Y_B & Y_C \end{matrix} \qquad (13.14)$$

Here is why it cannot be done. The winning combinations satisfy the following equations:

$$X_A X_B X_C = -1, \qquad (13.15)$$

$$X_A Y_B Y_C = 1, \quad Y_A X_B Y_C = 1, \quad Y_A Y_B X_C = 1. \qquad (13.16)$$

The product of the left-hand sides of Eqs. (13.16) equals $X_A X_B X_C$. The product of their right-hand sides equals $+1$. Obviously, these three equations cannot be satisfied as long as Eq. (13.15) holds. Pre-agreed answers offer no fail-safe strategy. And yet there is such a strategy [Greenberger *et al.* (1989)].

13.5.2 *A fail-safe strategy*

Here goes: Andy, Bob, and Charles prepare three spin-1/2 particles in a specific manner. Each player keeps one particle with him. When asked for the value of X, he will measure the x component of the spin of his particle, and when asked for the value of Y, he will measure the y component. His answer will be $+1$ or -1 according as his outcome is positive or negative. Proceeding in this way, the players are sure to win.

The three-particle state in question is

$$\| \Psi \rangle\!\rangle = \frac{1}{\sqrt{2}} |z_+, z_+, z_+\rangle - \frac{1}{\sqrt{2}} |z_-, z_-, z_-\rangle . \qquad (13.17)$$

Problem 13.9. *Using Eqs. (12.22) and (12.24), show that*

$$\| \Psi \rangle\!\rangle = \frac{1}{2} \Big(|x_+, x_+, x_-\rangle + |x_+, x_-, x_+\rangle + |x_-, x_+, x_+\rangle + |x_-, x_-, x_-\rangle \Big).$$

Each term contains x_- an odd number of times. Consequently, whenever the x components of the three spins are measured, the product of the outcomes will be negative.

Problem 13.10. *Using Eqs. (12.22–12.25), show that*

$$\| \Psi \rangle\!\rangle = \frac{1}{2} \Big(|x_+, y_+, y_+\rangle + |x_+, y_-, y_-\rangle + |x_-, y_+, y_-\rangle + |x_-, y_-, y_+\rangle \Big).$$

Each term contains an even number of minus subscripts. Moreover, since Eq. (13.17) is symmetric under permutations of the three spins, this holds whenever the spin state of one particle is written in terms of $|x_\pm\rangle$ and the spin states of the two others are written in terms of $|y_\pm\rangle$. Hence whenever the x component of one spin and the y components of the two other spins are measured, the product of the outcomes will be positive.

Can the outcomes obtained in these measurements reveal pre-existent values? For the same reason that Andy, Bob, and Charles could not use pre-agreed answers to ensure a win, the answer is negative.

By the time Greenberger *et al.* (1989) published their paper, it was all but taken for granted that the contradictions between quantum mechanics and "elements of physical reality" are essentially statistical. Bell-type inequalities are violated by the statistics of measurements performed on ensembles of systems associated with identical states. The observables represented by the operators in Kochen–Specker type arrays cannot all be measured together. When Greenberger *et al.* showed, by using an entangled state of *three* particles, that one can dispose of noncontextual hidden variables through a *single* false prediction, it caused quite a stir.

Here is how one can arrive at such a prediction: Suppose that three particles are associated with the state (13.17), that the values (13.16) pre-exist, and that each equals +1 or −1. We can use the first Eq. (13.16) to conclude that $X_A = Y_B Y_C$, we can use the second to conclude that $X_B = Y_A Y_C$, and we can use the third to conclude that $X_C = Y_A Y_B$. Based on these conclusions we predict that the product $X_A X_B X_C$ will come out equal to $Y_B Y_C Y_A Y_C Y_A Y_B = (Y_A)^2 (Y_B)^2 (Y_C)^2 = 1$. Yet if we measure this product, we invariably obtain the value −1.

The first GHZ-type experiment was performed by Bouwmeester *et al.* (1999). Needless to say, it was in agreement with the predictions of quantum mechanics.

13.6 Uses and abuses of counterfactual reasoning

As we just saw, a fruitful source of error in quantum mechanics is the illegitimate use of counterfactual reasoning. As another example, consider the following two-particle state [Hardy (1993); Mermin (1994)]:

$$\| \Psi \rangle\rangle = \sqrt{\frac{3}{8}} \, |A_R, A_G\rangle + \sqrt{\frac{3}{8}} \, |A_G, A_R\rangle - \frac{1}{2} \, |A_G, A_G\rangle. \qquad (13.18)$$

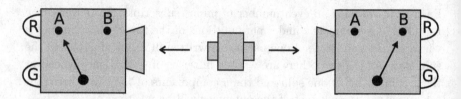

Fig. 13.2   Setup illustrating an illegitimate use of counterfactuals.

The setup is similar to that discussed in Sec. 13.1, except that the two apparatuses have two modes of working rather than three (Fig. 13.2). In mode A, each apparatus indicates either of two possible outcomes, $|A_R\rangle\langle A_R|$ or $|A_G\rangle\langle A_G|$. The vectors $|A_R\rangle$ and $|A_G\rangle$ thus form a basis in a 2-dimensional vector space.

Problem 13.11. $\|\,\Psi\,\rangle\!\rangle$ *is a unit vector in the direct product of the individual vector spaces.*

Problem 13.12. *The following vectors form another basis in the same space of states:*

$$|B_G\rangle = \sqrt{\frac{3}{5}}\,|A_G\rangle + \sqrt{\frac{2}{5}}\,|A_R\rangle, \quad |B_R\rangle = -\sqrt{\frac{2}{5}}\,|A_G\rangle + \sqrt{\frac{3}{5}}\,|A_R\rangle\,.$$

In mode B, each apparatus indicates either $|B_R\rangle\langle B_R|$ or $|B_G\rangle\langle B_G|$.

Problem 13.13. $(*)$ *Let* $\hat{\mathbf{A}}$ *and* $\hat{\mathbf{B}}$ *be the operators corresponding to the observables that each apparatus can measure, and let their eigenvalues be* ± 1 : $\hat{\mathbf{A}}|A_R\rangle = |A_R\rangle$, $\hat{\mathbf{A}}|A_G\rangle = -|A_G\rangle$, $\hat{\mathbf{B}}|B_R\rangle = |B_R\rangle$, $\hat{\mathbf{B}}|B_G\rangle = -|B_G\rangle$. *Write down the matrices for the two operators in the basis made up of* $|A_R\rangle$ *and* $|A_G\rangle$. *Show that they do not commute.*

Let us calculate some probabilities:

$$\langle\!\langle A_R, A_R \,\|\, \Psi \,\rangle\!\rangle = 0 \tag{13.19}$$

$$\langle\!\langle A_G, B_G \,\|\, \Psi \,\rangle\!\rangle = \sqrt{\frac{3}{5}}\langle\!\langle A_G, A_G \,\|\, \Psi \,\rangle\!\rangle + \sqrt{\frac{2}{5}}\langle\!\langle A_G, A_R \,\|\, \Psi \,\rangle\!\rangle$$

$$= -\frac{1}{2}\sqrt{\frac{3}{5}} + \sqrt{\frac{3}{8}}\sqrt{\frac{2}{5}} = 0\,. \tag{13.20}$$

In the same way one finds that $\langle\!\langle B_G, A_G \,\|\, \Psi \,\rangle\!\rangle = 0$. Thus if both apparatuses work in mode A, two red lights are never seen, and if they work in different modes, two green lights are never seen.

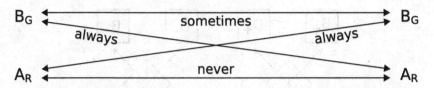

Fig. 13.3 The arrows illustrate the correlations between outcomes obtained by the measuring devices in Fig. 13.2.

Problem 13.14.

$$\langle\!\langle B_G, B_G \,\|\, \Psi \rangle\!\rangle = 3/10 \,. \tag{13.21}$$

If both apparatuses work in mode B, two green lights are seen with a probability of 9%. We may therefore assume that we have made the experiment with both apparatuses in mode B, and that both lights turned green.

What if, in the same run of the experiment, the first apparatus had been set to work in mode A? In this case its indicator light would have shown red. This is a valid counterfactual. We can check that every time the apparatuses work in mode B and one light is red, the other light will be green. For the same reason the following counterfactual is valid: if, in the same run of the experiment, the *second* apparatus had been set to work in mode A, its indicator light would have shown red. What we cannot do is invoke both these valid counterfactuals and draw the following conclusion: if, in the same run of the experiment, both apparatuses had worked in mode A, both lights would have shown red. Since two red lights are never seen if both apparatuses work in mode A (Fig. 13.3), this conclusion is obviously false.

The reason why is it generally illegitimate to combine legitimate counterfactuals is that their implicit "other things being equal" clauses are violated. In this particular case, the first counterfactual is valid only if the second apparatus is assumed to work as it actually did, and the second counterfactual is valid only if the first apparatus is assumed to work as it actually did.

Here is another way of being led up the garden path. The setup consists of two interlocking interferometers of the kind discussed in Sec. 11.8 (Fig. 13.4). In the left interferometer positrons are used instead of photons, and in the right interferometer electrons are used. (As this is a *gedanken* experiment, we need not worry about the practical problem of constructing an interferometer for positrons.) There are four detectors, two (B_- and D_-) for the electrons and two (B_+ and D_+) for the positrons. The letters B and D stand for "bright" and "dark," respectively. The "dark" detectors

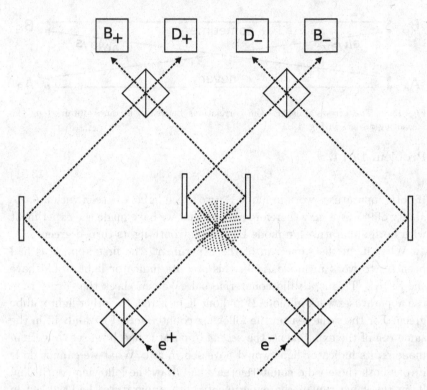

Fig. 13.4   Another experiment that lends itself to an illegitimate use of counterfactuals.

are so named because they never respond if a single electron or a single positron is launched. The fallacious argument goes like this:

An electron and a positron are simultaneously dispatched into their respective interferometers. If the electron alternatives interfere, D_- never responds. Hence if D_- clicks, the electron alternatives did not interfere. Something must have destroyed the interference between them, and this can only have been the positron in the other interferometer. But in order to be able to destroy the interference between the electron alternatives, the positron must have taken its inner (right) path.[1]

By the same token, if the positron alternatives interfere, D_+ never responds. Hence if D_+ clicks, the positron alternatives did not interfere. Something must have destroyed the interference between them, and this

---

[1]Although "destruction of interference" is a common phraseology, interference is not an object or state that can be destroyed. (Nor is it a physical mechanism or process, as was stressed in Sec. 5.3.) Interference is a feature that probability distributions display whenever they are calculated according to Rule B.

can only have been the electron in the other interferometer. But in order to be able to destroy the interference between the positron alternatives, the electron must have taken its inner (left) path.

Yet if both particles had taken their inner paths, they would have annihilated each other somewhere in the dotted region, and no detector would have responded. The assumption that both dark detectors click if both particles are dispatched together is thereby reduced to absurdity. The corresponding probability is therefore zero.

A quick calculation will show that this conclusion is wrong.

Let us call the electron alternatives O_- (for "outer path") and I_- (for "inner path"). The positron alternatives then are O_+ and I_+. If the two particles are launched together, there are four alternatives: $O_- \& O_+$, $O_- \& I_+$, $I_- \& O_+$, and $I_- \& I_+$. If the fourth alternative takes place, the two particles annihilate each other, and no detector responds. Hence only the first three alternatives contribute to the probability of "double dark detection" (both D's click). This probability is given by

$$p(D_-\&D_+) = |a(O_-\&O_+) + a(O_-\&I_+) + a(I_-\&O_+)|^2$$
$$= |a(O_-)\,a(O_+) + a(O_-)\,a(I_+) + a(I_-)\,a(O_+)|^2 . \quad (13.22)$$

The amplitudes $a(O_-)$, $a(I_-)$, $a(O_+)$, and $a(I_+)$ are again equal except that each contains an extra factor i for every reflection: $a(O_-) = Ai$, $a(O_+) = Ai$, $a(I_-) = Ai^3$, and $a(I_+) = Ai^3$. Thus,

$$p(D_-\&D_+) = |(Ai)(Ai) + (Ai)(Ai^3) + (Ai^3)(Ai)|^2 = |A^2|^2 . \quad (13.23)$$

Problem 13.15.

$$p(D_-\&B_+) = p(B_-\&D_+) = p(D_-\&D_+), \quad p(B_-\&B_+) = 9\,p(D_-\&D_+) .$$

Since the probability with which the two particles annihilate each other is $1/4$, the probability $p(D_-\&D_+) + p(D_-\&B_+) + p(B_-\&D_+) + p(B_-\&B_+)$, with which any two detectors respond, is $(1 + 1 + 1 + 9)|A^2|^2 = 3/4$. This tells us that $p(D_-\&D_+) = |A^2|^2 = 1/16$.

So where did the above argument go wrong? In the context of the Elitzur–Vaidman experiment (Sec. 11.8), it was legitimate to conclude from a response of the dark detector (D_2) that the bomb was present, and to conclude further that the alternative involving reflection by M_2 took place. In the present context, it is legitimate to conclude from a response of D_+ that the electron was present, for if only a positron had been launched, the probability of a response by D_+ would have been 0. But now there are two alternatives involving the positron, and nothing has happened that

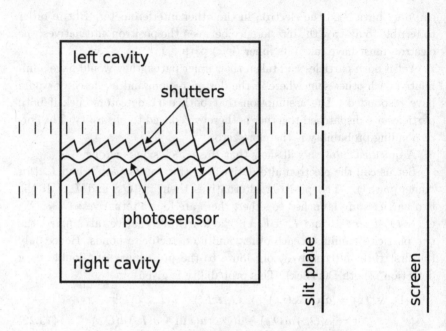

Fig. 13.5   The experiment of Englert, Scully, and Walther.

warrants the further conclusion that the electron has taken its inner path. By the same token, nothing has happened that warrants the conclusion that the positron has taken its inner path. What has once again led to the wrong conclusion is our deep-seated misconception that effects have to be locally produced.

13.7   The experiment of Englert, Scully, and Walther

Rule A (Sec. 5.1), you will recall, applies "if the intermediate measurements are made (or if it is possible to find out what their outcomes would have been if they had been made)," and Rule B applies "if the intermediate measurements are not made (and if it is impossible to find out what their outcomes would have been if they had been made)." The following two-slit experiment [Englert *et al.* (1994); Scully *et al.* (1991)] demonstrates the rationale behind the cryptic clauses in parentheses.

In this experiment atoms are used instead of electrons. All atoms are of the same type—Cesium-133, for example—and all start out in the same excited state $|e\rangle$. Placed in front of the slits are two initially separate microwave resonance cavities, each tuned to the energy difference ΔE between $|e\rangle$ and the ground state $|g\rangle$, and thus capable of holding photons[2] of energy ΔE. The design of each cavity moreover ensures that the probability with which an atom is found to emerge from it in the ground state equals unity—provided, as always, that the appropriate measurement is made.

13.7.1 *The experiment with shutters closed*

The two resonance cavities are separated from each other by a pair of electro-optical shutters, which remain closed for now. Atoms are launched, one at a time, with nothing to predict the particular cavity through which any given atom will pass. (Before an atom is launched, the photon left behind by the previous atom is "removed": the possibility of detecting it no longer exists.) Each atom leaves a mark on the screen. How will the marks be distributed?

Focus on a single atom, after it has hit the screen but before the photon is removed. This is a situation in which it is possible to find out what the outcome of an intermediate measurement would have been if it had been made. The intermediate measurement, had it been made, would have determined the slit taken by the atom. The reason why we can find out what its outcome would have been is a strict correlation between its outcome and the cavity containing the photon. If the atom were found to emerge from the left slit, the probability of detecting (and absorbing) a photon in the left cavity would be 1, and if the atom were found to emerge from the right slit, the probability of detecting (and absorbing) a photon in the right cavity would be 1. Conversely, if a photon is detected in the left (right) cavity, a measurement of the slit taken by the atom would have indicated that the atom has taken the left (right) slit.

Thus Rule A applies. Let us color the marks: those made by atoms that left a photon in the left cavity *green*, and those made by atoms that left a photon in the right cavity *red*. The dotted curve in Fig. 13.6 gives the distribution of the green marks, the dashed curve that of the red marks. The solid curve is the sum of the two distributions. The green marks are distributed as we expect from atoms that went through the left slit (L), and the red marks are distributed as we expect from atoms that went through the right slit (R). (Compare Fig. 13.6 with Fig. 5.2.)

---

[2]In what sense can a cavity hold a photon? In precisely this conditional sense: if a (100% efficient) photodetector were inserted into the cavity, it would detect a photon with probability 1.

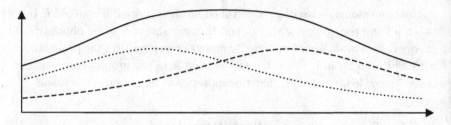

Fig. 13.6   Distribution of marks if the experiment is done with closed shutters.

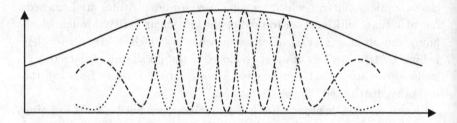

Fig. 13.7   Distribution of marks if the experiment is done with open shutters.

13.7.2   *The experiment with shutters opened*

Situated between the shutters there is a photosensor. If the shutters are opened before the photon is removed, and if the sensor is 100% efficient, quantum mechanics predicts that the sensor will absorb the energy ΔE with probability 1/2. Since we now have a single cavity instead of two, information about the slit taken by the photon is no longer available. (It has become customary to say that the information has been "erased." What has actually been "erased," however, is merely the possibility of obtaining the information.) Does this mean that Rule B now applies? If this experiment is done with sufficiently many atoms, will the overall distribution of marks exhibit interference fringes?

The answer has to be negative, for the measurement involving the photon is made *after* the atom has hit the screen. The decision about which measurement to perform—to determine the cavity that held the photon or to determine the behavior of the photosensor upon opening the shutters—comes too late to affect the overall distribution of marks.

But we now have another way of coloring the marks: *yellow* if the photosensor responds, *blue* if it fails to respond. Quantum mechanics predicts that the yellow marks will exhibit the same interference pattern as electrons in a two-slit experiment under the conditions stipulated by Rule B

(Fig. 5.3). Because the overall distribution of marks is the same in both versions of the experiment, the blue marks will exhibit the complementary interference pattern, having maxima where the other has minima and *vice versa*. The dotted curve in Fig. 13.7 gives the distribution of yellow marks, the dashed curve that of blue marks. The solid curve—the sum of the two distributions—is the same as in Fig. 13.6.

All "yellow" atoms and all "blue" atoms have something in common, but as their respective behaviors lack classical counterparts, we have no ready name for it. We may say that the "yellow" atoms went through the slits *in phase*, while the "blue" atoms went through the slits *out of phase*. We use these phrases for the following reason. If there is to be a·maximum at the center of the screen, the phases of the amplitudes associated with the alternatives "through L" and "through R" must differ by an even multiple of π—the alternatives must be "in phase." And if there is to be a minimum instead, the amplitudes must differ by an odd multiple of π—the alternatives must be "out of phase."

13.7.3 *Influencing the past*

The "green" atoms, we noted, behave like atoms that went through L, while the "red" atoms behave like atoms that went through R. Likewise, the "yellow" atoms behave like atoms that went through the slits in phase (inasmuch as they display the corresponding interference pattern), while the "blue" atoms behave like atoms that went through the slits out of phase. Cannot we conclude from this that the "green" atoms *actually* went through L, that the "red" atoms *actually* went through R, that the "yellow" atoms *actually* went through the slits in phase, and that the "blue" atoms *actually* went through the slits out of phase? After all, if it looks like a duck, swims like a duck, and quacks like a duck, then it probably is a duck. The problem with this conclusion is that it seems to imply the possibility of influencing the past.

If the experimenters determine the cavity that held the photon, they learn through which slit the corresponding atom went. If they open the shutters and observe whether or not the sensor responds, they learn how the atom went through the slits—in phase or out of phase. They cannot make the atom go through L or through R, yet by doing the former experiment, they can make sure that it went through a *single* slit—either L or R. Nor can they make an atom go through the slits in phase or out of phase, yet by doing the latter experiment they can make sure that it when through *both*

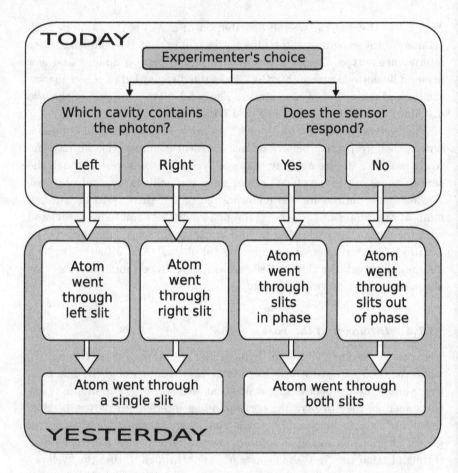

Fig. 13.8   Logical flow chart for the experiment of Englert, Scully, and Walther.

slits.[3] And since they can choose between the two experiments *after* the atom has made its mark on the screen, they can, by their choice, contribute to determine the atom's past behavior (Fig. 13.8).

Before rejecting the assumptions that lead to this conclusion, we need to be clear about what they amount to. Saying that a "green" atom went through L cannot mean that this very atom would also have gone through L if the cavity containing the photon emitted by it had not been ascertained. For if this cavity had not been ascertained, the experimenters could have checked *how* the atom went through both slits (in phase or out of phase),

---

[3]For more on the meaning of "both" see Sec. 18.1.2.

and regardless of the outcome they would have found that it went through both slits.

Rather, a "green" atom went through L only because this is what was indicated by a measurement. An atom goes through a particular slit only if the appropriate measurement is made, and it goes through L only if this is the outcome. It goes through both slits with a particular phase relation only if a different measurement is made, and it goes through both slits in phase only if that is the outcome. If neither measurement is made, it does not go through the left slit and it does not go through the right slit and it does not go through both slits in phase and it does not go through both slits out of phase. To paraphrase a well-known conclusion by John Wheeler,[4] no property (or behavior) is a possessed (occurrent) property (behavior) unless it is a measured (or indicated) property (behavior). Quantum-mechanical measurements do not merely *reveal* pre-existent properties or behaviors that occurred independently of measurements; they *create* their outcomes.

Seen in this light, the possibility of partially determining an atom's past behavior no longer seems preposterous. If, as a general rule, the properties of the quantum world exist only if, and only to the extent that, they are measured,[5] then the behavior of a quantum system at an earlier time can depend on a measurement performed at a later time, and thus also on a decision taken at a later time. Nothing like the backward-in-time causation found in some fantasy novels, which *changes* the past, is suggested. By choosing to perform either experiment, the experimenters cause no change in the atom's past behavior. Rather, they contribute to create or determine its past behavior. The world has exactly one history. Its state at any given time is what it is; it cannot be changed. But it can be what it is because of a measurement performed at a later time.[6]

But still, if no behavior is an occurrent behavior unless it is a measured or indicated behavior, we have reasons to doubt that the "green" atoms *actually* went through L. Strictly speaking, the conclusion that is warranted by the detection of a photon in the left cavity is this: the corresponding atom would have been found taking the left slit if a *direct* measurement of the slit taken by the atom had been made. But here we need to take into account that no measurement is direct. Measurement outcomes are inferred

---

[4] "No elementary phenomenon is a phenomenon until it is a registered (observed) phenomenon" [Wheeler (1983)].

[5] This issue will be conclusively addressed in Sec. 16.3.

[6] In fact, property-indicating events occur quite generally *after* the time of possession of the indicated property, albeit usually a fairly short time after.

from pointer positions, digital displays, detector clicks, computer printouts, and so forth. If we had no right to infer the position of a particle from such an indicator—if we could only conclude *ad infinitum* that a particle would be found with certainty if the appropriate measurement were made—then it would be impossible to measure the position of a particle, or anything else for that matter. But if we do have the right to infer the position of a particle from the click of a detector, then we also have the right to infer the slit taken by an atom from the click of a photosensor (if one is placed in each cavity).

13.8 Time-symmetric probability assignments

The Born rule (8.32) is generally interpreted asymmetrically with respect to time: the probability $p(\hat{\mathbf{P}}_v) = |\langle v|u\rangle|^2$ is assigned to the possible outcome $\hat{\mathbf{P}}_v = |v\rangle\langle v|$ of a measurement performed at a later time t_2 on the basis of an actual outcome $|u\rangle\langle u|$ of a measurement performed at an earlier time t_1. Given the temporally asymmetric character of human experience, it is obvious why we prefer this interpretation. Yet a time-reversed interpretation is just as legitimate; we can also use the Born rule to retrodict the probabilities of the possible outcomes of an earlier measurement on the basis of the actual outcome of a later measurement.

The difference between these two uses of the Born rule diminishes if we think in terms of the ensembles needed to (approximately) measure Born probabilities. Such ensembles can be postselected as well as preselected. To preselect an ensemble is to take into account only those instances of the measurement performed at t_1 that yield the particular outcome $\hat{\mathbf{P}}_u = |u\rangle\langle u|$. The preselected ensemble—an ensemble of identically *prepared* systems— serves to measure, as relative frequencies, the probabilities of the possible outcomes of the measurement performed at t_2. To postselect an ensemble is to take into account only those instances of the measurement performed at t_2 that yield the particular outcome $\hat{\mathbf{P}}_v = |v\rangle\langle v|$. The postselected ensemble—an ensemble of identically *retropared* systems—serves to measure, as relative frequencies, the probabilities of the possible outcomes of the measurement performed at t_1. (Since Born probabilities are conditional on the actual outcomes on the basis of which they are assigned, the "prepared" probabilities of the possible outcomes of a measurement can obviously differ from the "retropared" probabilities of the possible outcomes of the same measurement.)

Quantum mechanics even allows us to assign probabilities symmetrically with respect to time, on the basis of *both* earlier *and* later outcomes.

Suppose that three measurements are performed at the respective times $t_1 < t_2 < t_3$, that the measurement at t_1 yields the outcome u (represented by the projector $|u\rangle\langle u|$), and that the measurement at t_3 yields the outcome w (represented by the projector $|w\rangle\langle w|$). We can calculate the probability $p(v|w, u)$ with which the outcome v (represented by the projector $|v\rangle\langle v|$) is obtained at t_2 taking both data into account.

To measure this probability as a relative frequency, we use an ensemble that is both pre- and postselected (i.e., an ensemble of systems that are identically retropared as well as identically prepared). To create the appropriate ensemble, we take into account (i) only those instances of the measurement performed at t_1 that yield u and (ii) only those instances of the measurement performed at t_3 that yield w. In other words, we discard all runs in which the first measurement yields an outcome different from u and the third measurement yields an outcome different from w.

To calculate $p(v|w, u)$, we start from Eq. (1.8) but include in each term a condition c:

$$p(b|a,c) = \frac{p(a,b|c)}{p(a|c)} \, . \qquad (13.24)$$

In the present context this reads

$$p(v|w,u) = \frac{p(w,v|u)}{p_V(w|u)} \, . \qquad (13.25)$$

$p(w,v|u)$ is the probability of obtaining both v and w by the respective measurements at t_2 and t_3, on the condition that u is obtained at t_1. It is given by the product $p(w|v)\,p(v|u)$ of two Born probabilities. $p_V(w|u)$ is the probability of obtaining w at t_3 given that u is obtained at t_1 *and* given that the measurement with the possible outcome v is made at t_2. The subscript V stands for the observable that is measured at t_2. Since we calculate the probability with which this measurement yields v, we must use $p_V(w|u)$ rather than the Born probability $p(w|u) = |\langle w|u\rangle|^2$, which is applicable only if *no* measurement is made between t_1 and t_3. As the intermediate measurement is assumed to be made, Rule A applies, so $p_V(w|u)$ is given by

$$p_V(w|u) = \sum_{i=1}^{n} p(w, v_i|u) \, , \qquad (13.26)$$

where the values v_i, $i = 1, 2, \ldots, n$, are the possible outcomes of a measurement of V. Setting $v = v_k$, we arrive at the *ABL rule* [Aharonov, Bergmann, and Lebowitz (1964)]:

$$p(v_k | w, u) = \frac{|\langle w | v_k \rangle \langle v_k | u \rangle|^2}{\sum_{i=1}^{n} |\langle w | v_i \rangle \langle v_i | u \rangle|^2}. \tag{13.27}$$

13.8.1 *A three-hole experiment*

It will be instructive to apply the ABL rule to the following setup. This features a plate with three holes in it; let us call them A, B, and C. In front of the plate and equidistant from the holes there is a particle source (say, an electron gun G), and behind the plate, again equidistant from the holes, there is a particle detector D. Finally, interposed between C and D there is a device that causes a phase shift by π. An electron emerging from the holes is thus prepared in a way that can be "described" by the vector

$$|1\rangle = \frac{1}{\sqrt{3}} \left(|A\rangle + |B\rangle + |C\rangle \right), \tag{13.28}$$

and it is retropared in a way that can be "described" by the vector

$$|2\rangle = \frac{1}{\sqrt{3}} \left(|A\rangle + |B\rangle - |C\rangle \right). \tag{13.29}$$

The factor $e^{i\pi} = -1$ in front of $|C\rangle$ takes account of the phase shifting device.

We now add an apparatus M_A that can indicate whether or not a particle went through A. The two possible outcomes of this measurement are represented by $\hat{\mathbf{P}}_A = |A\rangle\langle A|$ ("through A") and $\hat{\mathbf{P}}_{A'} = \hat{\mathbf{P}}_B + \hat{\mathbf{P}}_C$ ("not through A"). The probability that a particle launched at G and detected at D is found to have gone through A is thus

$$p(A|2, 1) = \frac{|\langle 2 | \hat{\mathbf{P}}_A | 1 \rangle|^2}{|\langle 2 | \hat{\mathbf{P}}_A | 1 \rangle|^2 + |\langle 2 | \hat{\mathbf{P}}_{A'} | 1 \rangle|^2}. \tag{13.30}$$

Problem 13.16. $|\langle 2 | \hat{\mathbf{P}}_{A'} | 1 \rangle|^2 = 0$.

Thus $p(A|2, 1) = 1$. With this setup, a particle launched at G and detected at D is certain to be found taking hole A.

If, instead of M_A, we use an apparatus M_B that can indicate whether or not a particle went through B, we similarly find that $p(B|2, 1) = 1$. With this setup, a particle launched at G and detected at D is certain to be found taking hole B.

Finally we use an apparatus that can indicate through which of the three holes a particle went. The three possible outcomes of this measurement are represented by $\hat{\mathbf{P}}_A$, $\hat{\mathbf{P}}_B$, and $\hat{\mathbf{P}}_C$. The probability that a particle launched at G and detected at D is found taking hole A now equals

$$p(A|2,1) = \frac{|\langle 2|\hat{\mathbf{P}}_A|1\rangle|^2}{|\langle 2|\hat{\mathbf{P}}_A|1\rangle|^2 + |\langle 2|\hat{\mathbf{P}}_B|1\rangle|^2 + |\langle 2|\hat{\mathbf{P}}_C|1\rangle|^2} = \frac{1}{3}. \tag{13.31}$$

Time-symmetric probability assignments thus are in general *contextual*. The probability with which a particle launched at G and detected at D is found to go through A depends on the possible outcomes of the intermediate measurement. If the only possible outcome other than A is A', then $p(A|2,1) = 1$, and if the possible outcomes other than A are B and C, then $p(A|2,1) = 1/3$.[7]

---

[7]Although the present section does not fall under the heading of the present chapter, its inclusion is warranted by the contextuality that time-symmetric probability assignments share with probability assignments to outcomes of measurements performed on entangled systems.

Chapter 14

Quantum statistics

14.1 Scattering billiard balls

Two billiard balls coast toward each other with equal speed, then collide. Initially the balls move in opposite directions, so their momenta add up to zero. Because momentum is conserved, the total momentum remains zero, so after the collision the balls still move with equal speed in opposite directions. If the collision is perfectly head-on, each ball simply reverses its direction of motion. But suppose that the collision is uncontrollably somewhat off-center, so that each ball veers from its original direction of motion by some angle α. We cannot predict whether the balls will scatter at right angles (plus/minus some small angle ϵ), but we can estimate the probability with which they will do so.

There are two ways in which the balls can scatter at right angles. If we call the incoming balls N and S (suggesting that they move northward and southward, respectively), we may call the outgoing balls E and W. If either ball is as likely to be scattered eastward as westward, the corresponding probabilities are equal:

$$p(N{\to}W, S{\to}E) = p(N{\to}E, S{\to}W) \stackrel{\mathrm{Def}}{=} p. \tag{14.1}$$

The probability $p_\perp$ with which either possibility happens, no matter which, is the sum of these two probabilities:

$$p_\perp = p(N{\to}W, S{\to}E) + p(N{\to}E, S{\to}W) = 2p. \tag{14.2}$$

14.2 Scattering particles

We now replace the billiard balls with particles (Fig. 14.1). We again assume that the total momentum is zero, and that the scattering is elastic;

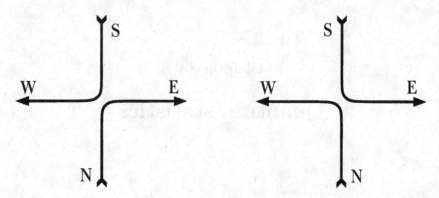

Fig. 14.1   The two possibilities that contribute to the probability with which two particles scatter elastically at right angles. The lines do not represent trajectories; they merely indicate possible identities between the incoming and outgoing particles.

in a particle context this means that no particles are created or annihilated in the process and that no type swapping takes place. If the incoming particles (and hence the outgoing ones) are of different types, then it is possible to learn which outgoing particle is identical with which incoming particle, so Rule A applies:

$$p_\perp = |A(N{\to}W, S{\to}E)|^2 + |A(N{\to}E, S{\to}W)|^2. \tag{14.3}$$

This is again the sum of two probabilities. If there are no preferred directions (due to external forces, particle spins, and such), the two probabilities are equal,

$$|A(N{\to}W, S{\to}E)|^2 = |A(N{\to}E, S{\to}W)|^2 \stackrel{\text{Def}}{=} p, \tag{14.4}$$

and we again have that $p_\perp = 2p$.

But now suppose that the conditions stipulated by Rule B are met. This rules out, among other things, that the incoming particles are of different types. We then have that

$$p_\perp = |A(N{\to}W, S{\to}E) + A(N{\to}E, S{\to}W)|^2. \tag{14.5}$$

Equation (14.4) allows the amplitudes to differ by a phase factor:

$$A \stackrel{\text{Def}}{=} A(N{\to}W, S{\to}E) = e^{i\phi} A(N{\to}E, S{\to}W). \tag{14.6}$$

Interchanging either the incoming or the outgoing particles, so that the alternative $N{\to}E, S{\to}W$ turns into the alternative $N{\to}W, S{\to}E$, causes the multiplication of $A(N{\to}E, S{\to}W)$ by $e^{i\phi}$. Accordingly, interchanging either the incoming or the outgoing particles *twice*, or interchange *both* the incoming particles and the outgoing particles, corresponds to a multiplication of $A(N{\to}E, S{\to}W)$ by $(e^{i\phi})^2$. But this simply restores the original alternative. It follows that $(e^{i\phi})^2 = 1$, and this leaves $e^{i\phi} = \pm 1$ as the only possibilities.

Nature makes use of both possibilities. Every particle is either a *boson* or a *fermion*. Exchanging two indistinguishable bosons leaves the amplitudes unchanged,

$$A(N{\to}W, S{\to}E) = A(N{\to}E, S{\to}W), \qquad (14.7)$$

whereas exchanging two indistinguishable fermions causes a change of sign,

$$A(N{\to}W, S{\to}E) = -A(N{\to}E, S{\to}W). \qquad (14.8)$$

For bosons, therefore,

$$p_\perp = |A + A|^2 = 4|A|^2 = 4p, \qquad (14.9)$$

whereas for fermions

$$p_\perp = |A - A|^2 = 0. \qquad (14.10)$$

Two indistinguishable bosons are *twice* as likely to scatter at right angles as two bosons that carry identity tags of some sort, while the probability with which two indistinguishable fermions scatter at right angles equals zero.

Can we nevertheless assume that there is an answer to the question "Which outgoing particle is identical with which incoming particle?" We cannot, for it this question had an answer, what actually happened would correspond to either of the alternatives of Fig. 14.1, and in this case we would have $p_\perp = 2p$ instead of either $p_\perp = 4p$ or $p_\perp = 0$.

14.2.1 *Indistinguishable macroscopic objects?*

Could the identities of macroscopic objects, such as billiard balls, get "mixed up" in this way?

For one thing, macroscopic objects are so large and/or complex that the likelihood with which two such objects are truly indistinguishable is virtually zero. For another, while it is relatively easy to adequately isolate two particles from the rest of the world, isolating sufficiently large or complex objects is virtually impossible. Even if the cosmic microwave background radiation were the only source of photons, such objects would continually absorb and emit photons with wavelengths in the millimeter range, and this would make it possible to pinpoint each object's location to within a few millimeters or less. If the objects are larger than that, then it is possible to trace their respective trajectories with sufficient precision to know which is which.

14.3 Symmetrization

Suppose that we have two bosons, and that their respective states are $|A\rangle$ and $|B\rangle$. And suppose that these bosons are fully characterized by their respective states. If we write the state of the composite system as $|A\rangle \otimes |B\rangle$, or simply $|AB\rangle$, we add to the physically warranted distinction between "the boson associated with $|A\rangle$" and "the boson associated with $|B\rangle$" the physically unwarranted distinction between the "first" or "left" boson and the "second" or "right" boson.

If $|A\rangle$ and $|B\rangle$ are orthogonal states, we can eliminate the physically unwarranted distinction by associating the composite system with the symmetric state

$$\frac{|AB\rangle + |BA\rangle}{\sqrt{2}}, \tag{14.11}$$

which is invariant under the interchange of A and B.

Problem 14.1. *If each of the orthogonal vectors $|A\rangle$ and $|B\rangle$ is a unit vector, then so is (14.11).*

Let $|A\rangle$, $|B\rangle$, $|C\rangle$, ... be mutually orthogonal boson states. Properly symmetrized multiple-boson states are obtained by adding all distinct permutations and dividing by the square root of their number. The symmetrized three-boson states are thus of the form

$$|AAA\rangle, \qquad \frac{1}{\sqrt{3}}\Big(|AAB\rangle + |ABA\rangle + |BAA\rangle\Big),$$

$$\frac{1}{\sqrt{6}}\Big(|ABC\rangle + |ACB\rangle + |BAC\rangle + |BCA\rangle + |CAB\rangle + |CBA\rangle\Big).$$

14.4 Bosons are gregarious

If n bosons have been found in possession of the same complete set of properties X and one boson has been found in possession of a different complete set of properties Y, what is the probability of finding all $n + 1$

bosons in possession of X? (A complete set of properties is what a complete measurement yields.) The initial and final states are, respectively,

$$|i\rangle = \frac{1}{\sqrt{n+1}}\Big(|\overbrace{X\dots X}^{n}Y\rangle + \cdots + |\overbrace{X\dots X}^{n-m}Y\overbrace{X\dots X}^{m}\rangle + \cdots + |Y\overbrace{X\dots X}^{n}\rangle\Big)$$

$$|f\rangle = |\overbrace{X\dots X}^{n+1}\rangle$$

The transition amplitude thus is

$$\frac{1}{\sqrt{n+1}}\Big(\langle\overbrace{X\dots X}^{n+1}|\overbrace{X\dots X}^{n}Y\rangle + \cdots + \langle\overbrace{X\dots X}^{n+1}|Y\overbrace{X\dots X}^{n}\rangle\Big). \quad (14.12)$$

Since the $n+1$ terms in brackets are equal to $\langle X|X\rangle^{n}\langle X|Y\rangle$, this amplitude is equal to

$$\frac{n+1}{\sqrt{n+1}}\langle\overbrace{X\dots X}^{n+1}|\overbrace{X\dots X}^{n}Y\rangle = \sqrt{n+1}\,\langle\overbrace{X\dots X}^{n+1}|\overbrace{X\dots X}^{n}Y\rangle. \quad (14.13)$$

The wanted probability thus is $(n+1)\,|\langle\overbrace{X\dots X}^{n+1}|\overbrace{X\dots X}^{n}Y\rangle|^{2}$.

If we were dealing with distinguishable ("tagged") bosons (in which case X and Y would not be complete sets of properties), the initial state would be the non-symmetrized vector

$$|\overbrace{X\dots X}^{n}Y\rangle,$$

and the transition probability would be $|\langle\overbrace{X\dots X}^{n+1}|\overbrace{X\dots X}^{n}Y\rangle|^{2}$.

The upshot: if n bosons of a given type have been found in possession of the same complete set of properties X and one boson of the same type has been found in possession of a different complete set of properties Y, the probability of finding all $n+1$ bosons in possession of X is $n+1$ times as large as it would be if the bosons were distinguishable.

14.5 Fermions are solitary

Fermion amplitudes change sign each time states are swapped between particles. Fermion states must therefore be anti-symmetrized. Since no anti-symmetric state can be formed of two identical fermion states, the probability of finding two fermions in possession of the same complete set of properties is zero. This is the content of the *exclusion principle* originally formulated by Wolfgang Pauli.

14.6   Quantum coins and quantum dice

We refer to the statistics of identical bosons as *Bose–Einstein* (BE) statistics, to that of identical fermions as *Fermi–Dirac* (FD) statistics, and to that of distinguishable particles as *Maxwell–Boltzmann* (MB) statistics.

As we all know, the probability with which a toss of a pair of MB coins results in two heads, $p_{MB}(HH)$, equals $p(H) \times p(H) = 1/4$. Since MB coins are distinguishable, there are two ways in which one comes up heads and one comes up tails, so the odds for this to happen are given by the sum of two probabilities, $p_{MB}(HT) + p_{MB}(TH) = 1/2$.

For a pair of FD coins, there is a single possible outcome, one heads and one tails, so the odds for this to happen equals unity. What about BE coins? The probability of tossing the first H is $p(H) = 1/2$. The probability of subsequently tossing another H is twice the probability of subsequently tossing a T, $p(H|H) = 2p(T|H)$, for the factor $n + 1$ equals 2 in this case. Because we must have that $p(T|H) + p(H|H) = 1$, we conclude that $p(T|H) = 1/3$ and $p(H|H) = 2/3$. The joint probability $p(HH)$—not to be confused with the conditional probability $p(H|H)$—equals $p(H|H) p(H) = (2/3)(1/2) = 1/3$. Similarly we find that $p(TT) = 1/3$. And since there is now only one way of obtaining one H and one T, the probability for this to happen is $1 - (1/3) - (1/3) = 1/3$. Note that the possible outcomes are again equiprobable. Only in the case of MB coins there are four possibilities while in the case of BE coins there are three.

Problem 14.2. (∗) *When tossing a pair of MB dice, the probability of obtaining the sum of 12, $p_{MB}(12)$, equals $1/6 \times 1/6 = 1/36$. There are two ways of obtaining the sum of 11, so $p_{MB}(11)$ is twice that. By the same token, $p_{MB}(10) = 3/36$, $p_{MB}(9) = 4/36$, and so on. What are the corresponding probabilities when tossing (i) a pair of BE dice and (ii) a pair of FD dice?*

The outcomes of tosses of MB dice are uncorrelated. The probabilities assigned to the possible outcomes of one toss are independent of the outcome of another toss. In particular, the probability of obtaining a 6 is $1/6$ regardless of how many dice already have come up 6. The outcomes of tossing BE dice or FD dice, on the other hand, are correlated. For BE dice, the probability of obtaining a 6 increases with the number of dice already showing a 6, while for FD dice the probability of obtaining a 6 is zero unless no other die is showing a 6.

What mechanism or process could possibly explain the correlations that obtain between indistinguishable particles? Misner *et al.* (1973) have answered this question succinctly: "No acceptable explanation for the miraculous identity of particles of the same type has ever been put forward. That identity must be regarded, not as a triviality, but as a central mystery of physics."

14.7 Measuring Sirius

The telescopic image of a pointlike light source is a disk whose diameter depends inversely on the telescope's aperture. For the largest telescopes, the angles subtended by the nearest or largest fixed stars are of the order of the diameter of this disk. At least this was true in 1956, when R. Hanbury Brown and R.Q. Twiss (1956) (HBT) showed how the angular diameter of a fixed star could be measured. HBT proposed to use two telescopes pointed at the same star. By fitting the telescopes with photomultipliers, they argued, the star's angular diameter could be deduced from correlations between variations of the photon detection rates through the telescopes.

The arguments of HBT were greeted with considerable disbelief, and various theoretical and experimental efforts were made to disprove them. Photons, it was thought, are emitted one at a time and detected one at a time. A photon is detected by either of the photomultipliers—M_1 or M_2. No photon is detected by both M_1 and M_2. So how could the photon counts be correlated? It was even thought that if HBT were correct, quantum mechanics was in need of thorough revision.

In 1965, an intensity interferometer using two paraboloid mirrors with a diameter of 6.5 m and a variable baseline (from 10 to 188 meters) was completed in Australia, and by the end of the decade, HBT had measured the angular diameters of more than twenty stars with remarkable accuracy, confounding those who had argued that their interferometer could not work [Hanbury Brown *et al.* (1967); Hanbury Brown (1968)].

The best way of demonstrating the error of HBT's critics is to go ahead and calculate. We first assume that there are two light sources. Let $a(y|x)$ be the amplitude associated with what we are used to calling "the emission of a photon" at $x = A$ or $x = B$ (the positions of the light sources) and "its absorption" at $y = 1$ or $y = 2$ (the positions of the telescopes). Let r_{yx} stand for the distance between x and y. The absolute value of $a(y|x)$ is inverse proportional to r_{yx}. Let the corresponding proportionality factor

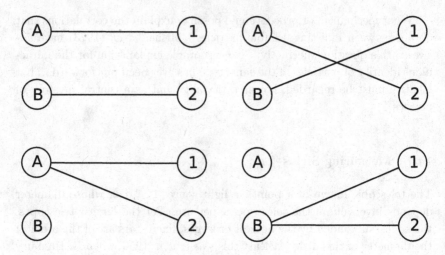

Fig. 14.2   Above: two indistinguishable alternatives; below: two distinguishable alternatives.

be α for the source at A and β for the source at B. The phase of $a(y|x)$ is proportional to r_{yx}. We may assume that the corresponding proportionality factor is the same wavenumber $k = 2\pi/\lambda$ for both sources. (Each telescope is fitted with a narrow bandpass filter, so that λ has a well-defined value.) Because the differences between the absolute values of the four amplitudes are minute as compared to the absolute values themselves, we may use

$$a(1|A) = \frac{\alpha}{R}e^{ikr_{1A}}, \quad a(1|B) = \frac{\beta}{R}e^{ikr_{1B}},$$
$$a(2|A) = \frac{\alpha}{R}e^{ikr_{2A}}, \quad a(2|B) = \frac{\beta}{R}e^{ikr_{2B}}.$$

R is defined by Fig. 14.3.

The simultaneous detection at 1 and 2 of photons emitted at A or B can come about in four ways. Two of the four alternatives—those with photons emitted by different sources as indicated in the upper half of Fig. 14.2—are indistinguishable, and two—those with photons emitted by the same source as indicated in the lower half of Fig. 14.2—are distinguishable, both from each other and from the indistinguishable alternatives. (By measurements performed on the sources at A and at B, it is in principle possible to learn how many photons were emitted by each source.) The amplitudes of the indistinguishable alternatives are

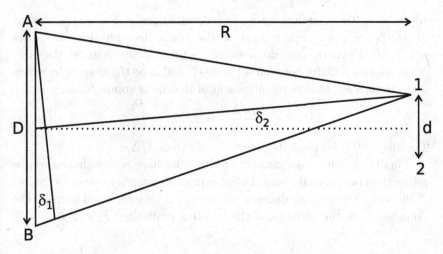

Fig. 14.3 Schematic of the parameters used in Hanbury Brown and Twiss's calculation of the angular diameter of a star.

$$a(1|A)\, a(2|B) = \frac{\alpha\beta}{R^2} e^{ik(r_{1A}+r_{2B})},$$

$$a(1|B)\, a(2|A) = \frac{\alpha\beta}{R^2} e^{ik(r_{1B}+r_{2A})}.$$

Because the alternatives are indistinguishable, we must add their amplitudes. The corresponding probability is therefore given by

$$p_I = \frac{\alpha^2\beta^2}{R^4} \left(e^{ik(r_{1A}+r_{2B})} + e^{ik(r_{1B}+r_{2A})} \right) \left(e^{-ik(r_{1A}+r_{2B})} + e^{-ik(r_{1B}+r_{2A})} \right)$$

$$= \frac{\alpha^2\beta^2}{R^4} \left(2 + e^{ik(r_{1A}+r_{2B}-r_{1B}-r_{2A})} + e^{ik(r_{1B}+r_{2A}-r_{1A}-r_{2B})} \right)$$

$$= \frac{2\alpha^2\beta^2}{R^4} \left(1 + \cos\left[k(r_{1A} + r_{2B} - r_{1B} - r_{2A}) \right] \right). \tag{14.14}$$

To this we have to add the probabilities of detecting photons from the same source. The probability of simultaneously detecting two photons thus is

$$P = p_I + |a(1|A)\, a(2|A)|^2 + |a(1|B)\, a(2|B)|^2 = p_I + \frac{\alpha^4}{R^4} + \frac{\beta^4}{R^4}$$

$$= \frac{1}{R^4} \left[\alpha^4 + \beta^4 + 2\alpha^2\beta^2 \left(1 + \cos\left[k(r_{1A} + r_{2B} - r_{1B} - r_{2A}) \right] \right) \right].$$

The angles δ_1 and δ_2 in what follows are defined by Fig. 14.3. Because $R \gg D$, where D is the distance between A and B, we have that

$$\delta_1 \approx \frac{r_{1B} - r_{1A}}{D} = \frac{r_{2A} - r_{2B}}{D} \approx \delta_2 \approx \frac{d}{2R},$$

where d is the distance between the telescopes. For the argument $(2\pi/\lambda)(r_{1A} + r_{2B} - r_{1B} - r_{2A})$ of the cosine we can therefore write $(2\pi/\lambda)\,dD/R$. As the distance between the telescopes is changed, the value of the cosine oscillates between $+1$ and -1. Let Δ be the amount by which d has to increase so that the argument of the cosine completes a cycle:

$$\frac{2\pi\,\Delta D}{\lambda R} = 2\pi. \qquad (14.15)$$

The angle D/R between the sources thus equals λ/Δ.

Finally we must take into account that we have a disk-shaped source rather than two pointlike ones. Doing so involves integration over the radius of the disk. The angular diameter of the star is then obtained in much the same way from the variation of the resulting probability P with Δ.

Chapter 15

Relativistic particles

15.1 The Klein–Gordon equation

Formally the complete Schrödinger equation (7.23) can be obtained by the following steps. Start with the non-relativistic relation between kinetic energy and kinetic momentum, $E_K = \mathbf{p}_K^2/2m$. Write this in terms of the total energy E, the total momentum $\mathbf{p}$, the potential energy $E_P = qV(\mathbf{r}, t)$, and the potential momentum $\mathbf{p}_P = (q/c)\,\mathbf{A}(\mathbf{r}, t)$ (Sec. 9.3):

$$E - qV = \frac{1}{2m}\left(\mathbf{p} - \frac{q}{c}\mathbf{A}\right)^2. \tag{15.1}$$

Multiply both sides from the right by $\psi(\mathbf{r}, t)$, and replace the total energy and the total momentum by the respective operators $\hat{E} = i\hbar\,\partial/\partial t$ and $\hat{\mathbf{p}} = -i\hbar\,\partial/\partial \mathbf{r}$ (Sec. 11.6):

$$\left(i\hbar\frac{\partial}{\partial t} - qV\right)\psi = \frac{1}{2m}\left(\frac{\hbar}{i}\frac{\partial}{\partial \mathbf{r}} - \frac{q}{c}\mathbf{A}\right)^2 \psi. \tag{15.2}$$

When we obtained the non-relativistic expressions for E_K and $\mathbf{p}_K$ from the relativistic expressions (9.19) and (9.20), we found that the kinetic parts of the relativistic expressions satisfy the relation (9.21). Using units in which $c = 1$, as is customary in the literature on relativistic quantum mechanics, this relation is

$$E^2 - \mathbf{p}^2 = m^2.$$

If we start with this relation and follow the same steps, we arrive at the relativistic version of the Schrödinger equation, which is known as the *Klein–Gordon equation* (although it was Schrödinger who considered it first):

$$\left[\left(i\hbar\frac{\partial}{\partial t} - qV\right)^2 - \left(\frac{\hbar}{i}\frac{\partial}{\partial \mathbf{r}} - q\mathbf{A}\right)^2\right]\psi = m^2\psi. \tag{15.3}$$

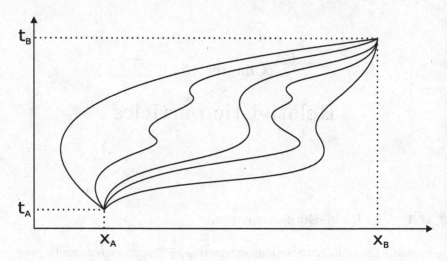

Fig. 15.1   Some of the paths over which we sum when calculating a non-relativistic particle propagator.

15.2   Antiparticles

As we noted in Sec. 4.1, any function of the form $\psi(\mathbf{r}, t) = A\, e^{(i/\hbar)(-Et + \mathbf{p}\cdot\mathbf{r})}$ satisfying the relation $E = \mathbf{p}^2/2m$ is a solution of the free Schrödinger equation, and so is any linear combination of such functions. By the same token, any function of the same form satisfying the relation $E^2 - \mathbf{p}^2 = m^2$, or $E = \pm\sqrt{m^2 + \mathbf{p}^2}$, is a solution of the free Klein–Gordon equation, and so is any linear combination of such functions. What, then, are we to make of solutions that seem to have *negative* energy?

When we introduced the (non-relativistic) propagator in the form of a path integral (Sec. 5.4), the summation extended over all paths leading from an initial to a final spacetime point, provided that velocities remained finite (Fig. 15.1). In the relativistic theory, spacelike path segments are exponentially suppressed rather than explicitly excluded (Sec. 9.1). But if paths with spacelike segments are not explicitly excluded, then paths that make U–turns with respect to time are not excluded either. If the speed of light can be exceeded, there is nothing to prevent a particle from returning into the past.

Two such paths are shown in Fig. 15.2. Paths in the vicinity of the solid curve, containing as they do sizable spacelike segments, make no significant contribution to a particle's amplitude of propagation from A to B. The same cannot be said of paths in the vicinity of the dashed curve, inasmuch

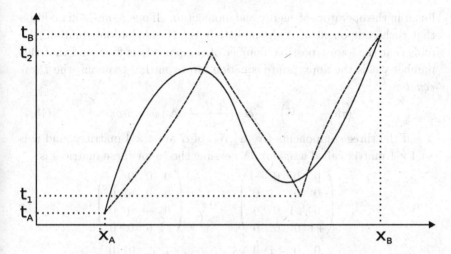

Fig. 15.2   Two paths that contribute to the particle propagator in a relativistic theory.

as this lacks spacelike segments. This path makes two U–turns (or rather, V–turns) with respect to time. According to a point of view advocated by Feynman (1949), a particle traveling in the positive time direction with negative energy is a particle traveling in the negative time direction with positive energy.[1] In more conventional terms, it is an *antiparticle*, which has the same mass and total spin as the corresponding particle but the *opposite charge*. Feynman's account of the dashed path in Fig. 15.2 is that a particle starts out at A, is scattered into a negative time direction at t_2, is scattered back into a positive time direction at t_1, and arrives at B. The conventional chronology, on the other hand, is that (i) a particle is launched at t_A, (ii) a particle-antiparticle pair is created at t_1, (iii) the particle launched at t_A and the antiparticle created at t_1 annihilate each other at t_2, and (iv) the particle created at t_1 is detected at t_B.

15.3   The Dirac equation

The utility of the Klein–Gordon equation is restricted to particles without spin. To arrive at the appropriate equation for a particle with spin, one needs to let the wave function have more than one component. This equation will therefore be a matrix equation. The simplest version is

---

[1] This chimes in with our observation in Sec. 6.6 that for antiparticles the sign of proper time differs from that of coordinate time.

linear in the operators of energy and momentum. If one requires in addition that each component of ψ should satisfy the Klein-Gordon equation, one finds that the lowest possible number of components is four, and that this number yields the appropriate equation for a spin-1/2 fermion—the *Dirac equation*:

$$\left(i\hbar\frac{\partial}{\partial t} - qV\right)\psi - \boldsymbol{\alpha}\cdot\left(\frac{\hbar}{i}\frac{\partial}{\partial\mathbf{r}} - q\mathbf{A}\right)\psi = \beta m\psi. \tag{15.4}$$

β and the three components $(\alpha_1, \alpha_2, \alpha_3)$ of $\boldsymbol{\alpha}$ are 4×4 matrices, and ψ is an 4×1 matrix called a *spinor*. A possible choice for these matrices is

$$\alpha_1 = \begin{pmatrix} 0 & 0 & 0 & +1 \\ 0 & 0 & +1 & 0 \\ 0 & +1 & 0 & 0 \\ +1 & 0 & 0 & 0 \end{pmatrix}, \quad \alpha_2 = \begin{pmatrix} 0 & 0 & 0 & -i \\ 0 & 0 & +i & 0 \\ 0 & -i & 0 & 0 \\ +i & 0 & 0 & 0 \end{pmatrix},$$

$$\alpha_3 = \begin{pmatrix} 0 & 0 & +1 & 0 \\ 0 & 0 & 0 & -1 \\ +1 & 0 & 0 & 0 \\ 0 & -1 & 0 & 0 \end{pmatrix}, \quad \beta = \begin{pmatrix} +1 & 0 & 0 & 0 \\ 0 & +1 & 0 & 0 \\ 0 & 0 & -1 & 0 \\ 0 & 0 & 0 & -1 \end{pmatrix}.$$

Note that the 2×2 matrices that make up the upper right and lower left quarters of the three α matrices are the Pauli spin matrices (12.26). The components of ψ correspond to the $2+2$ spin components of a spin-1/2 particle and its antiparticle. A more compact way of writing the Dirac equation (15.4) is

$$\left[i\gamma^k\left(\frac{\partial}{\partial x^k} + iqA_k\right) - m\right]\psi = 0, \tag{15.5}$$

where $\gamma^0 \overset{\text{Def}}{=} \beta$ and $\gamma^a \overset{\text{Def}}{=} \beta\alpha_a$ $(a=1,2,3)$. In the non-relativistic limit the Dirac equation reduces to the *Pauli equation*,

$$\left(i\hbar\frac{\partial}{\partial t} - qV\right)\psi = \frac{1}{2m}\left(\frac{\hbar}{i}\frac{\partial}{\partial\mathbf{r}} - \frac{q}{c}\mathbf{A}\right)^2\psi - \mu\,\boldsymbol{\sigma}\cdot\mathbf{B}\,\psi, \tag{15.6}$$

where ψ is a 2-component spinor, the three components of $\boldsymbol{\sigma}$ are the Pauli matrices, $\mathbf{B}$ is the magnetic field, and $\mu = q\hbar/2mc$ is the Bohr magneton introduced in Sec. 12.5.

15.4 The Euler–Lagrange equation

Consider now an action of the form

$$S_\phi[\mathcal{C}] = \int \mathcal{L}(\phi, \partial\phi/\partial x^k)\,d^4x, \tag{15.7}$$

where $\mathcal{C}$ is a path of a field ϕ in the field's configuration spacetime. We are interested in that path for which the action is stationary under infinitesimal variations of ϕ:

$$\delta S = \int \left[\frac{\partial \mathcal{L}}{\partial \phi} \delta \phi + \frac{\partial \mathcal{L}}{\partial (\partial \phi / \partial x^k)} \delta (\partial \phi / \partial x^k) \right] d^4 x = 0 . \qquad (15.8)$$

In general, the variation of ϕ takes place at every point in the spacetime region of integration, except at the temporal boundaries, where the initial and final field configurations are fixed. Integrating the second term in square brackets by parts and using $\delta(\partial \phi / \partial x^k) = \partial(\delta \phi)/\partial x^k$, we obtain

$$\delta S = \int \left[\frac{\partial \mathcal{L}}{\partial \phi} \delta \phi - \frac{\partial}{\partial x^k} \left(\frac{\partial \mathcal{L}}{\partial (\partial \phi / \partial x^k)} \right) \delta \phi + \frac{\partial}{\partial x^k} \left(\frac{\partial \mathcal{L}}{\partial (\partial \phi / \partial x^k)} \delta \phi \right) \right] d^4 x .$$

The last term can be converted into a surface integral over the boundary of the region of integration. As said, $\delta \phi$ vanishes at the temporal boundaries. If we consider only variations $\delta \phi$ that also vanish at the spatial boundary— for example because the field vanishes at spatial infinity—then this surface integral equals zero. And since $\delta S = 0$ holds for arbitrary variations $\delta \phi$, it implies that ϕ satisfies the *Euler–Lagrange equation*

$$\frac{\partial}{\partial x^k} \left(\frac{\partial \mathcal{L}}{\partial (\partial \phi / \partial x^k)} \right) - \frac{\partial \mathcal{L}}{\partial \phi} = 0 . \qquad (15.9)$$

If the Lagrangian contains several fields, there is one such equation for each field.

Problem 15.1. *The free Klein–Gordon equation (in units in which both c and $\hbar$ are equal to unity) is the Euler–Lagrange equation for a real field ϕ with Lagrangian*

$$\mathcal{L} = \frac{1}{2} \left[\left(\frac{\partial \phi}{\partial t} \right)^2 - \left(\frac{\partial \phi}{\partial \mathbf{r}} \right)^2 - m^2 \phi^2 \right] . \qquad (15.10)$$

The free Klein–Gordon equation is also the Euler–Lagrange equation for a complex field ϕ (which is equivalent to two real fields) with Lagrangian

$$\mathcal{L} = \left(\frac{\partial \phi}{\partial t} \right)^2 - \left(\frac{\partial \phi}{\partial \mathbf{r}} \right)^2 - m^2 \phi^2 . \qquad (15.11)$$

15.5   Noether's theorem

Consider, next, the infinitesimal form of a continuous deformation $\Delta\phi(\mathbf{r}, t)$ of the field ϕ,

$$\phi(\mathbf{r}, t) \to \phi'(\mathbf{r}, t) = \phi(\mathbf{r}, t) + \alpha\Delta\phi(\mathbf{r}, t), \qquad (15.12)$$

where α is an infinitesimal parameter. Here is how this deformation affects the Lagrangian $\mathcal{L}(\phi, \partial\phi/\partial x^k)$:

$$\alpha\Delta\mathcal{L} = \frac{\partial\mathcal{L}}{\partial\phi}(\alpha\Delta\phi) + \left(\frac{\partial\mathcal{L}}{\partial(\partial\phi/\partial x^k)}\right)\frac{\partial}{\partial x^k}(\alpha\Delta\phi)$$

$$= \alpha\frac{\partial}{\partial x^k}\left(\frac{\partial\mathcal{L}}{\partial(\partial\phi/\partial x^k)}\Delta\phi\right) + \alpha\left[\frac{\partial\mathcal{L}}{\partial\phi} - \frac{\partial}{\partial x^k}\left(\frac{\partial\mathcal{L}}{\partial(\partial\phi/\partial x^k)}\right)\right]\Delta\phi.$$

Because the last term vanishes in consequence of the Euler–Lagrange equation (15.9), we arrive at the following result: if $\mathcal{L}$ remains unaffected— $\Delta\mathcal{L} = 0$—by the infinitesimal deformation $\alpha\Delta\phi$, then the 4-current

$$j^k = \frac{\partial\mathcal{L}}{\partial(\partial\phi/\partial x^k)}\Delta\phi \qquad (15.13)$$

is conserved, i.e., it satisfies the equation of continuity $\partial j^k/\partial x^k = 0$. This exemplifies *Noether's theorem*, according to which a local conservation law is implied by the invariance of the Lagrangian under any continuous transformation of the fields on which it depends.

If, for example, $\mathcal{L}$ is invariant under spacetime translations, Noether's theorem implies the conservation of energy–momentum, and if $\mathcal{L}$ is invariant under rotations, it implies the conservation of angular momentum. It thereby allows us to derive expressions for the energy–momentum and the angular momentum of a closed system from the Lagrangian of the system.

It bears repetition, though, that an equation of continuity or a local conservation law does not warrant the conclusion that the conserved quantity is some localizable stuff moving about continuously. Local conservation laws are features of our calculational tools. As was pointed out in Sec. 10.7.1, the equation of continuity (10.37) for the components of energy–momentum (in flat spacetime) simply ensures that in all particle "collision" experiments, and regardless of the reference frame used, the total energy-momentum of the incoming particles equals the total energy–momentum of the outgoing particles.

If $\mathcal{L}(\phi, \partial\phi/\partial x^k)$ is invariant under the phase transformation $\phi \to e^{-i\alpha}\phi$, it is the system's charge that is conserved, and Noether's theorem allows us to obtain an expression for it. Applying the infinitesimal transformation

$$\alpha\Delta\phi = -i\alpha\phi, \quad \alpha\Delta\phi^* = i\alpha\phi^* \qquad (15.14)$$

to the Lagrangian (15.11), we obtain the locally conserved Noether current

$$j^k = i \left[\phi * \left(\frac{\partial \phi}{\partial x^k} \right) - \left(\frac{\partial \phi^*}{\partial x^k} \right) \phi \right] \tag{15.15}$$

associated with the (complex) Klein–Gordon equation. The charge density accordingly is

$$j^0 = \rho = i \left[\phi * \left(\frac{\partial \phi}{\partial t} \right) - \left(\frac{\partial \phi^*}{\partial t} \right) \phi \right]. \tag{15.16}$$

Here too there should be no uncertainty about the meaning of this local conservation law. Charge, like energy–momentum, is not a localizable stuff moving about continuously. The equation of continuity for charge simply ensures that in all particle "collision" experiments, and regardless of the reference frame used, the total charge of the incoming particles equals the total charge of the outgoing particles.

15.6 Scattering amplitudes

In non-relativistic physics, particle numbers are conserved. In a non-relativistic scattering experiment with N incoming particles, there will be N outgoing particles, and at any intermediate time, N particles would be found if the appropriate measurement were made. If there are several types of particles, this holds separately for each of them.

Not so in relativistic physics. Relativistic particles can leap into and out of existence in groups with zero total charge, and any group of particles can metamorphose into a different group of particles, provided that the overall charge (or charges) and the total energy–momentum stay the same. This means that we have to come to grips with a new kind of fuzziness—the fuzziness of particle numbers.

In consequence of this fuzziness, the amplitude $\langle \text{out} | S | \text{in} \rangle$ for a scattering event involving a group of incoming particles (specified by $|\text{in}\rangle$) and a group of outgoing particles (specified by $|\text{out}\rangle$) will be a series made up of terms that feature all possible particle metamorphoses in all possible combinations and arrangements. Formally this series can be obtained from a path integral of the following form:

$$\langle \text{out} | S | \text{in} \rangle = \int \mathcal{DC} \, e^{(i/\hbar) \int \mathcal{L}(\phi_a, \partial \phi_a / \partial x^k) \, d^4 x}. \tag{15.17}$$

$\mathcal{C}$ is now a path in the configuration spacetime of fields ϕ_a ($a = 1, 2, \ldots$) leading from the configuration specified by $|\text{in}\rangle$ (the quantum states of the incoming particles) to the configuration specified by |out⟩ (the quantum states of the outgoing particles). Because the fuzziness associated with individual particles has been taken into account by using a Lagrangian whose free part has the free wave equation for its Euler–Lagrange equation (Sec. 15.4), what is being summed over is alternatives that differ in the number of particles that are created and/or annihilated, the spacetime locations of the creation/annihilation events, and the (topologically) different ways in which these events can be connected by free-particle propagators.

In the case of fields associated with freely propagating particles there is nothing to be summed. If particles are to scatter each other, the Lagrangian must contain, apart from the term that yields the free field equations, an interaction term $\mathcal{L}_{\text{int}}$, which contains products of different fields or higher powers of a field than the second. Expanding $\exp[(i/\hbar)\int\mathcal{L}_{\text{int}}d^4x]$ in powers of the fields yields the desired series. Needless to say, this is only useful if the series converges rapidly enough, as it does for the electromagnetic and weak interactions.

15.7   QED

In SI units, the Lagrangian for quantum electrodynamics (QED) is

$$\mathcal{L} = \overline{\psi}\left[i\gamma^k\left(\frac{\partial}{\partial x^k} + iqA_k\right) - m\right]\psi - \frac{1}{4}F_{jk}F^{jk}, \tag{15.18}$$

where $\overline{\psi} \overset{\text{Def}}{=} \psi^\dagger\gamma^0$, $\psi^\dagger$ being the adjoint of ψ. We gather from this that $\mathcal{L}_{\text{int}} = -q\overline{\psi}\gamma^k A_k\psi$. The terms of the series for a given scattering amplitude can be represented by diagrams known as *Feynman diagrams*, according to rules known as *Feynman rules*, which encapsulate how diagrams are generated and what factors are associated with each diagram. The form of $\mathcal{L}_{\text{int}}$ indicates that QED diagrams are made of two types of lines—straight lines for fermions and wiggly lines for photons—and a single type of vertex: a point at which two fermion lines and one photon line meet.

15.8   A few words about renormalization

If the diagram in Fig. 15.3 were the only contribution to the electron–electron scattering amplitude, the two parameters appearing in the Lagrangian (15.18)—q and m—would be measurable and could be interpreted as the electron's charge and mass, respectively. There is, however, an

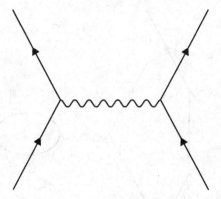

Fig. 15.3   Lowest-order graph contributing to the amplitude for electromagnetic scattering between electrons. The direction of the arrows (upward) signifies that the interacting particles are electrons rather than positrons. Because the mathematical expression represented by this graph contains, for each vertex, an integral over all possible spacetime locations, a vertex does not occupy any particular spacetime location.

infinite series of contributions (Fig. 15.4). We should therefore not be surprised if these parameters turned out to be physically meaningless, as in fact they do.

One way to obtain physical—which is to say, measurable—expressions is to first attend to another apparent evil: when the diagrams are evaluated by integrating over the 4-momenta of the (internal) lines rather than over the positions of the vertices, one finds that most of the diagrams are divergent, which is to say, infinite. This can be repaired by introducing a cutoff, a procedure known as *regularization*: one integrates only up to a large but finite energy–momentum. One thereby obtains an expression like the following, which depends on the "bare" charge q, on the cutoff Λ, and on the momenta of the incoming and outgoing particles, whose values we may summarily denote by K:

$$A = -iq + iCq^2 \ln \frac{\Lambda^2}{K} + 0(q^3)\,. \tag{15.19}$$

C is a finite constant and $0(q^3)$ contains only higher powers of q than the second.

Suppose now that the actual, physical charge—call it q_p—has been measured for a specific set of 4-momenta K_0. To second order in q, we then have

$$-iq_p = -iq + iCq^2 \ln \frac{\Lambda^2}{K_0} + 0(q^3)\,. \tag{15.20}$$

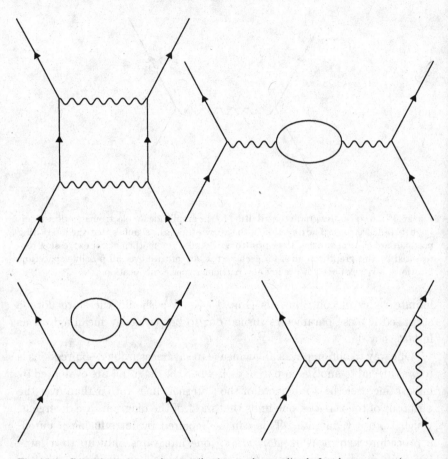

Fig. 15.4   Some 4-vertex graphs contributing to the amplitude for electromagnetic scattering between electrons.

This tell us that

$$-iq = -iq_p - iCq^2 \ln \frac{\Lambda^2}{K_0} + 0(q^3) = -iq_p - iCq_p^2 \ln \frac{\Lambda^2}{K_0} + 0(q_p^3). \quad (15.21)$$

The second equality is correct to the order of approximation indicated. Plug this into Eq. (15.19) to find that

$$
\begin{aligned}
A &= -iq_p - iCq_p^2 \ln \frac{\Lambda^2}{K_0} + iCq_p^2 \ln \frac{\Lambda^2}{K} + 0(q^3) \\
&= -iq_p + iCq_p^2 \ln \frac{K_0}{K} + 0(q^3).
\end{aligned} \quad (15.22)
$$

Lo and behold! Both the "bare" charge q and the cutoff Λ have disappeared. This little exercise illustrates that charges (as well as masses) are "running" parameters: depending on the Lagrangian, their physical values increase or decrease as the momentum scale at which experiments are performed increases.[2]

Only a few Lagrangians are nice enough to be *renormalizable*, in the sense that all experimental predictions can be made to converge by substituting physical parameters for a *finite* number of "bare" parameters in the manner indicated.

15.8.1 ... and about Feynman diagrams

How should we think about Feynman diagrams, and especially about their internal lines—those that correspond to neither incoming nor outgoing particles and are usually referred to as "virtual particles"? According to Zee (2003, pp. 53–57),

> Spacetime Feynman diagrams are literally pictures of what happened.... Feynman diagrams can be thought of simply as pictures in spacetime of the antics of particles, coming together, colliding and producing other particles, and so on.

Mattuck (1976) is more cautious:

> Because of the unphysical properties of Feynman diagrams, many writers do not give them any physical interpretation at all, but simply regard them as a mnemonic device for writing down any term in the perturbation expansion. However, the diagrams are so vividly 'physical looking', that it seems a bit extreme to completely reject any sort of physical interpretation whatsoever.... Therefore, we shall here adopt a compromise attitude, i.e., we will 'talk about' the diagrams as if they were physical, but remember that in reality they are only 'apparently physical' or '*quasi-physical*'.

Consider a two-electron scattering event. The momenta of the incoming particles are known, as are the momenta of the outgoing particles. Whatever happened in the meantime, during which nothing is measured, is anybody's guess. Zee would probably agree that whenever quantum mechanics requires us to sum over two or more alternatives, no single alternative can by itself be a picture of what actually happened. If indeed there is a sense in which alternatives represent what actually happened,

---

[2] The dependence of a particle's physical mass on a momentum scale in relativistic *quantum* physics should not be confused with the dependence of a particle's mass on its velocity, which is (quite unnecessarily) introduced in many textbooks on *classical* relativity.

this can only be that they all happened together. What this might mean will be spelled out in Part 3. It certainly has nothing to do with the magical invocation of "vacuum fluctuations," no matter how often this is repeated.[3]

Mattuck's plea for cognitive dissonance, on the other hand, is a recipe for philosophical disaster. Who has not heard the song and dance about a cloud of virtual photons and virtual particle-antiparticle pairs surrounding every real particle, which is routinely invoked in explanations of the dependence of physical masses and charges on the momentum scale at which they are measured? As long as this naive reification of quantum-mechanical probability algorithms continues, it should not surprise anyone that quantum mechanics keeps being vilified as unintelligible, absurd, or plain silly.[4]

15.9   Beyond QED

The well-tested quantum theories beyond QED are collectively known as the Standard Model. Wilczek (2001) went so far as to suggest that it be called simply "the theory of matter." It encompasses the theory of the strong nuclear interactions called quantum chromodynamics (QCD) and a partially unified theory of the weak and electromagnetic interactions—"partially" because it involves two symmetry groups rather than a single one. The findings of every high-energy physics experiment carried out to date are consistent with the Standard Model. While it does have real and/or perceived shortcomings, its proposed extensions have so far remained in the realm of pure speculation.

---

[3]Many Feynman diagrams contain so-called "vacuum parts," which are not connected by any sequence of lines and vertices to any incoming or outgoing particle. These are artifacts of the methods employed in generating the perturbation series. They are systematically canceled out in every actual calculation of a scattering amplitude. They certainly do not warrant the claim that "the vacuum in quantum field theory is a stormy sea of quantum fluctuations" [Zee (2003), p. 19].

[4]Another instance of cognitive dissonance was provided by Wilczek (1999) who, in one and the same article, declared (i) that electrons are "excitations of the same underlying ur-stuff, the electron field," which is "the primary reality," and (ii) that "we can only require—and generally we only obtain—sensible, finite answers when we ask questions that have direct, operational meaning." Questions about underlying ur-stuff and primary realities are obviously not of the kind that have direct, operational meaning.

15.9.1   QED revisited

Both QCD and the electroweak theory are generalizations of QED.

The Lagrangian (15.18) for QED is invariant under the gauge transformation (10.2–10.3):

$$\psi(t, \mathbf{r}) \to e^{iq\alpha(t, \mathbf{r})} \psi(t, \mathbf{r}), \tag{15.23}$$

$$A_k \to A_k - \frac{\partial \alpha}{\partial x_k} \quad \text{or} \quad V \to V - \frac{\partial \alpha}{\partial t}, \quad \mathbf{A} \to \mathbf{A} + \frac{\partial \alpha}{\partial \mathbf{r}}. \tag{15.24}$$

The invariance under the phase transformation (15.23) gives us the freedom to associate a complex plane with each spacetime point and to independently rotate the real and imaginary axes of each of these planes—as long as an infinitesimal difference in spacetime location corresponds to an infinitesimal difference in the angle of rotation. This is quite analogous to the freedom to associate an inertial system with each spacetime point and to independently Lorentz transform its axes. Both freedoms make room for a type of interaction. While the freedom to locally rotate the spacetime axes makes it possible to introduce the gravitational interaction, the freedom to locally rotate the axes of the complex plane makes it possible—in fact, *necessary*—to introduce the electromagnetic interaction, as the following will show.

Consider the Lagrangian that yields the free Dirac equation:

$$\mathcal{L} = \overline{\psi} \left(i\gamma^k \frac{\partial}{\partial x^k} - m \right) \psi. \tag{15.25}$$

If we want it to be invariant under (15.23), we must introduce a vector field A_k by substituting

$$\frac{\partial}{\partial x^k} \to \frac{\partial}{\partial x^k} + iqA_k \tag{15.26}$$

and we must require that A_k transform according to (15.24).[5] The need to add the term proportional to $F_{jk}F^{jk}$, which completes the QED Lagrangian (15.18), then arises for much the same reasons as it did in the classical context (not to mention the crucial requirement of renormalizability).

15.9.2   Groups

In mathematics, a *group* is a set $\mathcal{G}$ together with an operation that combines any two group elements a, b to form a third element $a \cdot b$. The set and the operation satisfy the following conditions:

---

[5]This substitution parallels the introduction of the covariant derivative in Sec. 9.6.4. A_k here plays the part of the connection coefficients (9.57).

(1) For all a, b in $\mathcal{G}$, $a \cdot b$ is also in $\mathcal{G}$.

(2) For all a, b, c in $\mathcal{G}$, $(a \cdot b) \cdot c = a \cdot (b \cdot c)$.

(3) $\mathcal{G}$ contains an identity element e such that $a = a \cdot e = e \cdot a$.

(4) Each a in $\mathcal{G}$ has an inverse b such that $a \cdot b = b \cdot a = e$.

Problem 15.2. *For any spacetime point $(t, \mathbf{r})$, the phase transformations (15.23) form a group.*

Because a phase factor can be thought of as a unitary 1×1 matrix, this group is called $U(1)$. As you will remember (Secs. 8.9 and 8.12.1), a linear operator is called unitary if (and only if) $\hat{\mathbf{U}}^\dagger \hat{\mathbf{U}} = \hat{\mathbf{I}}$ (Eq. 8.38), where $\hat{\mathbf{U}}^\dagger$ is the adjoint of $\hat{\mathbf{U}}$. If A_{ik} are the matrix components of a linear operator $\hat{\mathbf{A}}$, then the matrix components of $\hat{\mathbf{A}}^\dagger$ are A_{ki}^* (Sec. 12.1.2). For an element $U = e^{i\beta}$ of $U(1)$, Eq. (8.38) thus reduces to $U^*U = 1$.

If the elements of a group commute, so that $a \cdot b = b \cdot a$ for all a, b, the group is called *Abelian*. $U(1)$ is an example of such a group. The group of rotations in a 3-dimensional space is an example of a non-Abelian group: the product of two rotations generally depends on the order in which they are performed.

A group must be distinguished from its *representations*, which realize the group elements as matrices and the group operation as matrix multiplication (Sec. 12.1.2). The 2×2 matrices we encountered in Sec. 12.1 do not, in fact, belong to a representation of the group of rotations in 3-dimensional space. For, as you learned by doing Problem 12.2 and is also evident from Eq. (12.15), a rotation by 2π changes the sign of every vector, so that the smallest positive angle that yields the identity matrix is 4π rather than 2π. Those matrices belong to the fundamental or defining representation of $SU(2)$, the group of special unitary 2×2 matrices. "Special" here means that their determinants are equal to unity. $SU(2)$ is locally isomorphic to the rotation group, in the sense that for sufficiently small angles there is a one-to-one correspondence between the elements of the two groups and the operations in each group, but it covers the rotation group twice since to every element of the rotation group there correspond two elements of $SU(2)$.

15.9.3   *Generalizing QED*

Arguably the simplest and most straightforward generalization of QED is obtained by the following recipe. We again start with the Dirac Lagrangian (15.25), but instead of requiring its invariance under the local

$U(1)$ operation (15.23), we require its invariance under the local $SU(2)$ operation (12.31):

$$\psi(t, \mathbf{r}) \rightarrow e^{-i\,\boldsymbol{\theta}(t, \mathbf{r}) \cdot \mathbf{s}}\, \psi(t, \mathbf{r}). \qquad (15.27)$$

This means that, in addition to being a 4-component Dirac spinor on which the matrices γ^k act, ψ is now a 2-component vector on which the three components of $\mathbf{s}$ act. Each of these components—the *generators* of $SU(2)$—is a 2×2 matrix. To make the Lagrangian (15.25) invariant under the transformation (15.27), we must introduce a vector field $\mathbf{A}_k$ via a substitution analogous to (15.26), the difference being that each of the four spacetime components of the *gauge potential* $\mathbf{A}_k$ is now a 2×2 matrix that acts on the two vector components of ψ. Since each element of $SU(2)$ can be written as a linear combination of the group generators, we can introduce the components A_k^a and write $\mathbf{A}_k = \sum_{a=1}^{3} A_k^a s^a$. Furthermore we must require that the transformation (15.27) goes hand in hand with a transformation of $\mathbf{A}_k$ that is analogous to the transformation (15.24), albeit more involved due to the non-Abelian nature of $SU(2)$.

Finally we need to add a term analogous to that proportional to $F_{jk}F^{jk}$ in the Lagrangian for QED. This term has to be constructed out of the components A_k^a of the gauge potential, their first derivatives, and the *structure constants* of $SU(2)$, which are defined by (12.32) and given by the antisymmetric symbol (12.33). Here, too, the construction leaves little to the physicist's discretion.

A further generalization of QED is obtained by following the same recipe with $SU(3)$ in place of $SU(2)$. It takes eighteen real numbers to specify a complex 3×3 matrix. Unitarity reduces the number of independent parameters to nine (Eq. 12.12), and the required unity of the determinant reduces it further to eight. $SU(3)$ therefore has eight generators.

The three gauge groups $U(1)$, $SU(2)$, and $SU(3)$ are the very ones that feature in the Standard Model, the first two as ingredients of the unified theory of the electromagnetic and weak interactions, the third as the gauge group of of the strong interaction or QCD.

15.9.4   *QCD*

The $SU(3)$ symmetry of QCD requires the strongly interacting particles, which are known as *quarks*, to come in three varieties. These are characterized by their charges, which are collectively known as *colors*. The eight varieties of massless bosons that are said to "mediate" the strong force, called *gluons*, also carry color charges, owing to the non-Abelian nature of $SU(3)$.

The two most important characteristics of QCD are *asymptotic freedom* and *quark confinement*. The first means that the shorter the distances over which quarks interact, the more weakly they interact. The second means that only color-neutral combinations of quarks will ever appear in isolation. Both baryons and mesons are such combinations. Baryons—among them the proton and the neutron—are bound states of three quarks. Mesons are bound states of a quark and an antiquark.[6] Although it is still a hypothesis, quark confinement is widely believed to be true; it explains the consistent failure of searches for free quarks, and it has been demonstrated by an approximate calculation scheme that formulates QCD on a grid or lattice of spacetime points.

Studied at a sufficiently high resolution, the atomic nucleus thus presents itself as containing quarks that interact by exchanging gluons, while at a lower resolution it presents itself as containing protons and neutrons that interact by exchanging mesons. The force that binds the protons and neutrons in a nucleus turns out to be a residue of the strong force, comparable to the interactions between the permanent or induced dipole moments of electrically neutral atoms or molecules, which are residual manifestations of the electrostatic force between electrically charged particles.

15.9.5 *Electroweak interactions*

The construction of the electroweak Lagrangian is rather more involved. One of its salient features is that the left-handed leptons enter differently from the right-handed ones. So what are leptons and what is meant by their handedness?

The fundamental particles of the Standard Model are the quarks, which interact via the strong force, and the leptons, which do not. There are six types of *quark* or quark "flavor" (d, u, s, c, b, t) and six *leptons* (the electron e, the muon μ, the tauon τ, and three corresponding neutrinos ν_e, ν_μ, ν_τ). These particles group themselves into three *generations*. The first generation contains the d and u ("down" and "up") quarks, the electron,

---

[6]Here is a simple illustration of confinement: Imagine trying to separate the quark q from the antiquark $\bar{q}$ in a meson $q\bar{q}$. The energy required to do this grows with the distance between the two quarks. At some point the energy supplied is enough to create a new quark-antiquark pair, q' and $\bar{q}'$, so that you end up with two mesons, $q\bar{q}'$ and $q'\bar{q}$, instead of two separate quarks.

and the electron–neutrino; the second contains the s and c ("strange" and "charmed") quarks, the μ, and the ν_μ; the third contains the b and t ("bottom" and "top") quarks, the τ, and the ν_τ. Except for their masses, the particles of the second and third generations have the same properties, and thus interact in the same way, as the corresponding particles of the first generation. It will therefore suffice to consider only the first generation.[7]

As to the handedness of leptons, a spin-1/2 particle in a momentum eigenstate defines a spatial axis by the direction of the corresponding eigenvalue **p**. If a measurement of the particle's spin with respect to this axis yields "up" with probability 1, the particle is said to be *right-handed*, and if it yields "down" with probability 1, the particle is said to be *left-handed*.

The left-handed leptons (and therefore also the right-handed anti-leptons) interact via the weak force; the right-handed leptons do not "feel" this force. The lepton part of the Lagrangian therefore contains two terms, a "right-handed" one constructed so as to make it invariant only under local $U(1)$ transformations, and a "left-handed" one constructed so as to make it invariant under both local $U(1)$ transformations and local $SU(2)$ transformations. Thus there will be a $U(1)$ gauge field B_k and an $SU(2)$ gauge field C_k^a. Apart from the differential treatment of left-handed and right-handed leptons and the absence of a lepton mass term corresponding to $-\overline{\psi}m\psi$ in the Dirac Lagrangian (15.25), the construction of these parts follows the standard recipe. So does the construction of the pure gauge field part.

15.9.6 *Higgs mechanism*

But that's only the beginning, as it were. The bosons that are said to "mediate" the weak force are known to have large masses, yet the inclusion of an explicit mass term for the $SU(2)$ gauge field would destroy the gauge invariance and render the theory non-renormalizable. The answer to this conundrum is a procedure that, for brevity, is often referred to as "Higgs mechanism."

The first step of this procedure is to introduce a spin-0 field ϕ. The next step is to create a degeneracy of the vacuum state by adopting a particular combination of a term quartic in ϕ and a term quadratic in ϕ. (The vacuum state is the quantum state with the lowest possible energy. Saying that it is degenerate is the same as saying that it is not unique.) To

---

[7]We are here glossing over the mixing of flavors of different generations in the eletroweak Lagrangian.

again have a unique vacuum state, one defines new fields in terms of the ones so far present in the Lagrangian, and one "fixes the gauge" (i.e., chooses a specific gauge) by performing a specific gauge transformation. As a result, the $SU(2)$ symmetry is "spontaneously broken." (This is something of a misnomer, as fixing the gauge is a rather deliberate process.) The new fields are associated with three massive spin-1 bosons, the $W^\pm$ and the Z^0 (which are said to "mediate" the weak force), the massless photon, and the massive spin-0 Higgs boson (whose existence has not been experimentally confirmed at the time of this writing).

The electroweak Lagrangian also contains an interaction term for the leptons and the Higgs. When applied to this term, the Higgs mechanism gives rise to a mass term for the leptons. When quarks are included (again without a mass term in the original Lagrangian), they too end up having masses courtesy of the Higgs mechanism. Unsurprisingly, therefore, this procedure has been hailed as *explaining* why particles have mass. But does it meet what is expected of an explanation?

To keep our feet on the ground, we must not forget that what we are dealing with is essentially a mathematical tool for calculating scattering amplitudes. Given the numbers, types, and 4-momenta of the incoming particles, this puts us in a position to calculate the probabilities for all possible combinations of outgoing particles. What it does not furnish is a clue to the nature of the underlying physical mechanism or process—if at all there is such a thing. How then could it furnish a clue to the creation of mass, even if mass were some kind of stuff or thing, rather than a parameter of a calculational tool. All that is achieved by the Higgs mechanism—and this is feat enough—is the computability of scattering amplitudes involving weak interactions, by giving rise to mass terms without jeopardizing renormalizability.

PART 3
Making Sense

Chapter 16

Pitfalls

16.1 Standard axioms: A critique

There can be little doubt that the most effective way of introducing the *mathematical formalism* of quantum mechanics is the axiomatic approach. Philosophically, however, this has its dangers. Axioms are supposed to be clear and compelling. The standard axioms of quantum mechanics are neither. Because they lack a convincing physical motivation, students—but not only students—tend to accept them as ultimate encapsulations of the way things are.

The first standard axiom typically tells us that the state of a system S is (or is represented by) a normalized element $|v\rangle$ of a Hilbert space $\mathcal{H}_S$.[1] The next axiom usually states that observables are (or are represented by) self-adjoint linear operators on $\mathcal{H}_S$, and that the possible outcomes of a measurement of an observable $\hat{O}$ are the eigenvalues of $\hat{O}$.

Then comes an axiom (or a couple of axioms) concerning the (time) evolution of states. Between measurements (if not always), states are said to evolve according to unitary transformations, whereas at the time of a measurement, they are said to evolve (or appear to evolve) as stipulated by the projection postulate: if $\hat{O}$ is measured, the subsequent state of S is the eigenvector corresponding to the outcome, regardless of the previous state of S.

A further axiom stipulates that the states of composite systems are (or are represented by) vectors in the direct product of the Hilbert spaces of the component systems.

Finally there are a couple of axioms concerning probabilities. According to the first, if S is in the state $|v\rangle$, and if we do an experiment to see if it

---

[1] Weinberg (1996) is nearer the mark when he represents the state of S by a *ray* in $\mathcal{H}_S$.

has the property $|w\rangle\langle w|$, then the probability of a positive outcome is given by the Born rule $p = |\langle w|v\rangle|^2$. According to the second, the expectation value of an observable $\hat{O}$ in the state $|v\rangle$ is $\langle v|\hat{O}|v\rangle$.

There is much here that is perplexing if not simply wrong. To begin with, what is the physical meaning of saying that the state of a system is (or is represented by) a normalized vector (or else a ray) in a Hilbert space? The main reason why this question seems virtually unanswerable is that probabilities are introduced almost as an afterthought. It ought to be stated at the outset that the mathematical formalism of quantum mechanics is a probability calculus. It provides us with algorithms for calculating the .probabilities of measurement outcomes.

If both the phase space formalism of classical physics and the Hilbert space formalism of quantum physics are understood as tools for calculating the probabilities of measurement outcomes, the transition from a 0-dimensional point in a phase space to a 1-dimensional ray in a Hilbert space is readily understood as a straightforward way of making room for nontrivial probabilities (Sec. 8.4). Because the probabilities assigned by the points of a phase space are trivial, the classical formalism admits of an alternative interpretation: we may think of (classical) states as collections of possessed properties. Because the probabilities assigned by the rays of a Hilbert space are nontrivial, the quantum formalism does not admit of such an interpretation: we may not think of (quantum) states as collections of possessed properties.

Saying that the state of a quantum system is (or is represented by) a vector or a ray in a Hilbert space, is therefore seriously misleading. There are two kinds of things that can be represented by a vector or a ray in a Hilbert space: possible measurement outcomes and actual measurement outcomes. If a possible measurement outcome is represented by a vector or a ray in a Hilbert space, it is so *solely* for the purpose of calculating its probability. If an actual measurement outcome is represented by a vector or a ray in a Hilbert space, it is so *solely* for the purpose of assigning probabilities to the possible outcomes of whichever measurement is made next. Thus if $|v\rangle$ represents the outcome of a maximal test, and if $|w\rangle$ represents a possible outcome of the measurement that is made next, then the probability of that outcome is $|\langle w|v\rangle|^2$. (If the Hamiltonian is not zero, a unitary operator, taking care of the time difference between the two measurements, has to be sandwiched between $\langle w|$ and $|v\rangle$.)

It is essential to understand that any statement about a quantum system *between measurements* is "not even wrong" in Pauli's famous phrase, inasmuch as such a statement is neither verifiable nor falsifiable. This bears on the third axiom (or couple of axioms), according to which quantum states evolve (or appear to evolve) unitarily between measurements, which then implies that they "collapse" (or appear to do so) at the time of a measurement.

All that can safely be asserted about the time t on which a quantum state functionally depends is that it refers to the time of a measurement. If $|v(t)\rangle$ represents the outcome of a maximal test on the basis of which probabilities are assigned, then t is the time of this maximal test. If $|w(t)\rangle$ represents one of the possible outcomes of a measurement to which probabilities are assigned, then t is the time of this measurement. What cannot be asserted without metaphysically embroidering the axioms of quantum mechanics is that $|v(t)\rangle$ is (or represents) an instantaneous state of affairs of some kind, which evolves from earlier to later times. As Peres (1984) pointedly observed, "there is no interpolating wave function giving the 'state of the system' between measurements."

Again, what could be the physical meaning of saying that observables are (or are represented by) self-adjoint operators? We are left in the dark until we get to the last couple of axioms, at which point we learn that the expectation value of an observable $\hat{\mathbf{V}}$ in the state $|w\rangle$ is $\langle w|\hat{\mathbf{V}}|w\rangle$. As we have seen in Sec. 11.2, if (as a result of a previous measurement) the vector $|w\rangle$ is associated with S, and if we contemplate a measurement whose possible outcomes v_k are represented by the projectors $|v_k\rangle\langle v_k|$, then we can *define* a self-adjoint operator $\hat{\mathbf{V}}$ such that the expectation value $\sum_k v_k |\langle v_k|w\rangle|^2$ (i.e., the sum over the possible outcomes weighted by their Born probabilities) can be written as $\langle w|\hat{\mathbf{V}}|w\rangle$.

And finally, why would the state of a composite system be (represented by) a vector in the direct product of the Hilbert spaces of the component systems? Once again the answer is virtually self-evident (Sec. 13.2) if quantum states are seen for what they are—tools for assigning nontrivial probabilities to the possible outcomes of measurements.

16.2 The principle of evolution

Why is the fundamental theoretical framework of contemporary physics a *probability calculus*, and why are the events to which the probabilities are assigned *measurement outcomes*? It seems to me that all previous attempts to arrive at satisfactory answers to these vexed questions have foundered on two assumptions. The first is the notion that physics can be neatly divided into kinematics, which concerns the description of a system at an *instant* of time, and dynamics, which concerns the *evolution* of a system from earlier to later times. We may call this notion "the principle (or paradigm) of evolution."

By itself, the principle of evolution implies that time is infinitely or completely differentiated, so that we are permitted to think of it as a set of temporally unextended instants. In combination with the special theory of relativity—specifically, its frame-dependent stratification of spacetime into hyperplanes of constant time—the principle of evolution implies that space, too, is infinitely or completely differentiated, so that we are permitted to think of it as a set of (spatially) unextended points. We may call this "the principle (or paradigm) of complete spatiotemporal differentiation." As we shall see, both principles present serious obstacles to making sense of the quantum theory.

In keeping with the principle of evolution, the wave function $\psi(x, t)$ is usually awarded the primary status, while the propagator $\langle x_f, t_f | x_i, t_i \rangle$ is seen as playing a secondary role, which is to relate the wave function at different times according to

$$\psi(x_f, t_f) = \int dx_i \, \langle x_f, t_f | x_i, t_i \rangle \, \psi(x_i, t_i), \qquad (16.1)$$

notwithstanding that both wave function and propagator encapsulate the same information. Underlying this partiality towards the wave function is the idea that wave functions—and quantum states in general—are meaningful constructs even in the absence of measurements. If this were the case, measurements would merely contribute (or even merely appear to contribute) to the determination of quantum states, and quantum states would determine absolute probabilities rather than probabilities that are always conditional on measurement outcomes.[2]

The prevalence of these ideas can be traced back to two fortuitous cases of historical precedence: that of Schrödinger's "wave mechanics" over Feynman's propagator-based formulation of the theory, and that of Kolmogorov's (1950) formulation of probability theory over an axiomatic alternative developed by Rényi (1955, 1970). Every result of Kolmogorov's theory has a translation into Rényi's [Primas (2003)]. Yet whereas in

---

[2]The conditionality of quantum-mechanical probabilities—their dependence on actual measurement outcomes, in addition to the setup-specific boundary conditions under which they are assigned—has also been stressed by Primas (2003).

Kolmogorov's theory absolute probabilities have primacy over conditional ones, Rényi's theory is based entirely on conditional probabilities.

The principle of evolution is also at odds with the time-symmetry of the quantum-mechanical correlation laws. For we can use the Born rule to retrodict the probabilities of the possible outcomes of an earlier measurement on the basis of the actual outcome of a later measurement as well as to predict the probabilities of the possible outcomes of a later measurement on the basis of the actual outcome of an earlier measurement. As Sec. 13.8 has shown, quantum mechanics even allows us to assign probabilities symmetrically with respect to time, on the basis of both earlier and later outcomes. Positing an interpolating quantum state evolving from later to earlier measurements would therefore seem just as legitimate—or, rather, illegitimate—as positing an interpolating quantum state evolving from earlier to later measurements.

16.3   The eigenstate–eigenvalue link

There is a widely held if not always explicitly stated assumption, which for many has the status of an additional axiom. This is the so-called *eigenstate-eigenvalue link*, according to which a system "in" an eigenstate of the operator $\hat{\mathbf{B}}$ associated with an observable B possesses the corresponding eigenvalue even if no measurement of B is actually made. Because the time-dependence of a quantum state is a dependence on the time of a measurement, rather than the continuous time-dependence of an evolving state, we need to reject this assumption. All that $\hat{\mathbf{B}}|v(t)\rangle = b\,|v(t)\rangle$ implies is that a (successful) measurement of B made at the time t is certain to yield the outcome b. Probability 1 is not sufficient for "is" or "has."

If a system's being in an eigenstate of an observable (qua operator) is not sufficient for the possession, by the system or the observable, of the corresponding eigenvalue, then what is? In Chap. 13 we came across several experimental arrangements that warranted the following conclusion: measurements do not reveal pre-existent values; instead, they create their outcomes. If this is correct, then the only sufficient condition for the existence of a value $v(B)$ is a measurement of the observable B. Observables have values only if, only when, and only to the extent that they are measured. To *be* is to be *measured*.[3]

---

[3]This is the idea that Bohr tried to convey by stressing that out of relation to experimental arrangements the properties of quantum systems are undefined [Jammer (1974); Petersen (1968)].

There would appear to be a narrow escape route for the proponents of hidden variable theories: accept the contextuality of pre-existent values (Sec. 13.4). To many physicists and philosophers of science, this seems too great a price to pay for the rather modest benefits—essentially psychological, they would say—provided by hidden variables.

It seems to me that the attempt to salvage hidden variables through contextuality is not only costly but also self-defeating. If measurements create their outcomes, then it is an obvious possibility that the outcome of a measurement of an observable A depends on what other observables are measured together with A. But if we are concerned with pre-existent values, no reference to "measurement" should be made. All we are then entitled to say is that the value of b_2 in array (13.13), for instance, does not exist independently of the other values in this array. Pre-existent values only exist as sets. While as a member of a row, b_2 has the value -1, as a member of a column, it has the value $+1$. But how can one even speak of such sets if one is not allowed to refer to "measurement"? After all, membership in the same set is defined by being measured together.

Chapter 17

Interpretational strategy

For those who subscribe to the principle of evolution, the pivotal role played by measurements in all standard formulations of quantum mechanics is an embarrassment known as the "measurement problem." As an anonymous referee for a philosophy of science journal once put it to me, "to solve this problem means to design an interpretation in which measurement processes are not different in principle from ordinary physical interactions." But how can reducing measurements to "ordinary physical interactions" solve this problem, considering that quantum mechanics describes "ordinary physical interactions" in terms of correlations between the probabilities of the possible outcomes of *measurements* performed on the interacting systems? This kind of "solution" amounts to sweeping the problem under the rug.

In reality, to solve the measurement problem means to design an interpretation in which the central role played by measurements in standard axiomatizations of quantum mechanics is understood. And before it can be understood, one must acknowledge the obvious: that the formalism of quantum mechanics is a probability calculus, and that the events to which this assigns probabilities are measurement outcomes.

An algorithm for assigning probabilities to possible measurement outcomes on the basis of actual outcomes has two perfectly normal dependences. It depends continuously on the time of measurement: if this changes by a small amount, the assigned probabilities change by small amounts. And it depends discontinuously on the outcomes that constitute the assignment basis: if this changes by the inclusion of an outcome not previously taken into account, so do the assigned probabilities.

But think of a quantum state's dependence on time as the continuous time-dependence of an evolving state (rather than as a dependence on the time of a measurement), and you have two modes of evolution for the price of one:

(1) between measurements, a quantum state evolves according to a unitary transformation and thus continuously and predictably;

(2) at the time of a measurement, a quantum state generally "collapses": it changes (or appears to change) discontinuously and unpredictably into a state depending on the measurement's outcome.

Hence the mother of all quantum-theoretical pseudo-questions: why does a quantum state have (or appear to have) two modes of evolution? And hence the embarrassment. Getting rid of the pseudo-question is easy: we only have to recognize that the true number of modes of evolution is neither two nor one but zero. Getting rid of the embarrassment requires more work, for we still have two Rules for the price of one—those stated in Sec. 5.1 and derived in Sec. 8.13. Why *two* computational rules? What distinguishes this question from the above pseudo-question is that it has a straightforward answer:

Whenever quantum mechanics instructs us to add amplitudes rather than probabilities (i.e., whenever we are required to use Rule B rather than Rule A), the distinctions we make between the alternatives correspond to nothing in the actual world. They don't exist in the actual world. They exist solely in our minds.

This answer lies at the heart of the interpretational strategy adopted in the following pages. It does raise further questions, but it also makes it possible to answer them. In addition, it does not appeal to untestable metaphysical assumptions like hidden variables or evolving quantum states but refers directly to testable computational rules.

Chapter 18

Spatial aspects of the quantum world

18.1 The two-slit experiment revisited

Let us return to the two-slit experiment discussed in Sec. 5.2. This featured two alternatives:

- The electron went through the left slit (L).
- The electron went through the right slit (R).

Under the conditions stipulated by Rule A (Sec. 5.1), the probability of detection by a detector (located at) D is the sum of two probabilities, $p_L = |A_L|^2$ and $p_R = |A_R|^2$, where A_L and A_R are the amplitudes associated with the alternatives. This is consistent with the view that an electron detected at D went through either the left slit or the right slit.

Let us try to understand what happened, under the conditions stipulated by Rule B, when an electron is detected at D. Let us assume, to begin with, that

(1) each electron goes through a particular slit (either L or R),
(2) the behavior of electrons that go through a given slit does not depend on whether the other slit is open or shut.

If the first assumption is true, then the distribution of hits across the backdrop, when both slits are open, is given by

$$n(x) = n_L(x) + n_R(x), \tag{18.1}$$

where $n_L(x)$ and $n_R(x)$ are the respective distributions of hits from electrons that went through L and electrons that went through R. If the second assumption is true, then we can observe $n_L(x)$ by keeping the right slit shut, and we can observe $n_R(x)$ by keeping the left slit shut. What we observe when the right slit is shut is the dotted hump on the left side of Fig. 5.2, and what we observe when the left slit is shut is the dotted hump on the right side of this figure. If both assumptions are true, we thus expect to observe the sum of these two humps. But this is what we observe under the conditions stipulated by Rule A. What we observe under the conditions stipulated by Rule B is the interference pattern plotted in Fig. 5.3. At least one of the two assumptions is therefore false.

18.1.1   *Bohmian mechanics*

According to an interpretational strategy proposed by Bohm (1952), only the second assumption is false: all electrons follow well-defined paths, which wiggle in a peculiar manner and cluster at the backdrop so as to produce the observed distribution of hits.

What causes the wiggles? Bohmians explain this by positing the existence of a "pilot wave" that guides the electrons by exerting on them a force. If both slits are open, it passes through both slits. The secondary waves emanating from the slits interfere, with the result that the electrons are guided along wiggly paths.

The reason why, according to Bohmians, electrons emerging from the same source or the same slit arrive in different places, is that they start out in slightly different directions and/or with slightly different speeds. If we had precise knowledge of these initial values, we would be in a position to predict each electron's future motion with classical precision. Since quantum mechanics says that such knowledge cannot be had, Bohmians must say that, although well-defined electron paths and exact initial values exist, they are hidden from us. What they do not say is *why* these things are hidden from us, in spite of the fact that there is a simple answer to this question: they are hidden from us because they do not exist.

Bohmian mechanics is an extreme instantiation of the principle of evolution. It not only posits a wave function that evolves between measurements but also attributes to it the reality of a classical force acting on classical particles, in blithe disregard of the fact that the pilot wave associated with a physical system with n degrees of freedom propagates in an n-dimensional configuration space, which can be identified with physical space only if $n = 3$.[1]

---

[1] Another unpalatable feature that ought to be mentioned is that on this theory energy and momentum and spin and every particle property other than position are *contextual* [Albert (1992)].

18.1.2 The meaning of "both"

We are committed by our own interpretational strategy to conclude that the first assumption is wrong (in which case the second assumption is vacuous). Under the conditions stipulated by Rule B, each electron does in some sense pass through *both* slits. What does this mean? Saying that an electron went through both slits cannot be equivalent to the conjunction

"(the electron went through L) and (the electron went through R)."

To ascertain the truth of a conjunction, we must individually ascertain the truths of its components, yet we never find (i) that an electron launched at G and detected at D has taken the left slit and (ii) that the same electron has taken the right slit.

Nor can saying that an electron went through both slits mean that a part of the electron went through L and another part went through R. In fact, the question of parts does not arise. Analogous experiments have been performed with C_{60} molecules using a grating with 50 nm wide slits and a period of 100 nm [Arndt *et al.* (1999)]. The sixty carbon nuclei of C_{60} are arranged like the corners of soccer ball just 0.7 nm across. We do not picture parts of such a molecule as getting separated by many times 100 nm and then reassemble into a ball less than a nanometer across.

Saying that an electron went through both slits can only mean that it went through $L\&R$—the cutouts in the slit plate *considered as an undifferentiated whole*. Whenever Rule B applies, the distinction we make between L and R is a distinction that has no reality as far as the electron is concerned. In other words, the distinction between "the electron went through L" and "the electron went through R" is a distinction that "Nature does not make"—it corresponds to nothing in the actual world. The position at which the electron passed the slit plate is the entire undifferentiated region $L\&R$, or the entire undifferentiated union of two segments of a spatial plane. It is not any part or segment of $L\&R$, let alone a point in $L\&R$.

18.2 The importance of unperformed measurements

Take another look at Fig. 11.4. As stated in Sec. 11.8, what we see in these images is neither the nucleus nor the electron but the fuzzy relative position between the electron and the nucleus in various stationary states of atomic hydrogen. Nor do we see this fuzzy position "as it is." What we see is the plot of a position probability distribution, which depends on the outcomes of measurements determining the values of the atomic quantum numbers n, l, and m. This distribution both quantifies and defines the fuzzy relative position between the electron and the nucleus, by assigning probabilities counterfactually, to the possible outcomes of *unperformed* measurements.

To elucidate this point, I ask you to imagine a small bounded region V in the imaginary space of sharp positions relative to the proton, well inside the probability distribution associated with the electron's position relative to the proton. As long as this distribution is associated with the atom, the electron is neither inside V nor outside V. (If it were inside, the probability of finding it inside would be 1, and if it were outside, the probability of finding it inside would be 0, neither of which is the case.) If on the other hand we were to ascertain (by making the appropriate measurement) whether the electron was inside V or outside V, we would find it either inside V or outside V. We would change the fuzzy position we meant to describe, quantify, or define. Hence if we want to quantitatively describe a fuzzy observable, we must assume that a measurement is made, and if we do not want to change the observable in the process of describing it, we must assume that no measurement is made.[2] In other words, we must describe it by assigning probabilities to the possible outcomes of unperformed measurements.

The fact that fuzzy observables are quantified by assigning probabilities to the possible outcomes of (unperformed) measurements, goes a long way towards explaining why quantum mechanics is a probability calculus, and why measurements enjoy the special status that they do. It also shows that Bell's criticism was beside the point. "To restrict quantum mechanics to be exclusively about piddling laboratory operations is to betray the great enterprise," he wrote [Bell (1990)]. The unperformed measurements that are key to the quantitative description of fuzzy observables cannot be called "piddling laboratory operations," nor is the occurrence of measurements restricted to laboratories. *Any* event or state of affairs from which either the truth or the falsity of a proposition of the form "system S has the property P" (or "observable O has the value V") can be inferred, qualifies as a measurement.

---

[2]This is not a contradiction but the very meaning of a counterfactual statement.

18.3   Spatial distinctions: Relative and contingent

What kind of relation exists between an electron and a region V if the electron is neither inside V nor outside V? If, as appears to be the case, being inside and being outside are the only relations that can hold between an object's position and a region of space, then no kind of relation exists between the electron and V. In this case V simply does not exist as far as the electron is concerned. And since conceiving of a region V is tantamount to making the distinction between "inside V" and "outside V," we are led to conclude again that the distinction we make between "inside V" and "outside V" has no reality for the electron. The distinction we make between "the electron is inside V" and "the electron is outside V" corresponds to nothing in the actual world.

It follows that the reality of a spatial distinction is *relative*: the distinction we make between disjoint regions may exist for one object at one time and not exist for a different object at the same time or for the same object at a different time. To give an example, a device capable of indicating the slit taken by an electron may not always function as intended—no detector is 100% efficient. If it functioned as intended, a relation existed between each slit and the position of the electron at the time when it passed the slit plate. The propositions "the electron went through L" and "the electron went through R" are both in possession of truth values: one is true, the other is false. In this case the distinction between L and R has been real for the electron. If on the other hand the device failed to indicate the slit taken by an electron, neither proposition has a truth value, and no relation existed between the electron and either slit. In this case the distinction we make between L and R did not exist as far as this electron is concerned.

The reality of a spatial distinction is also *contingent*: whether the distinction we make between "inside V" and "outside V" is real for a given object O at a given time t depends on whether the proposition "O is in V at t" has a truth value (either "true" or "false"), and this in turn depends on whether either O's presence in V or O's absence from V (at the time t) is indicated by an actual event or state of affairs.

18.4   The importance of detectors

If the reality of spatial distinctions is relative and contingent, physical space cannot be something that by itself has parts. For if the regions defined by any conceivable partition were intrinsic to space, and therefore distinct by themselves, the distinctions we make between them would be real for every object in space. It follows that a detector is needed not only to indicate the presence of an object in its sensitive region R but also, and in the first place, to realize (make real) a region R, by realizing the distinction between being inside R and being outside R. It thereby makes the predicates "inside R" and "outside R" available for attribution.

And this bears generalization, not least because "in physics the only observations we must consider are position observations, if only the positions of instrument pointers" [Bell (1987), p. 166]. The measurement apparatus that is presupposed by every quantum-mechanical probability assignment is needed not only for the purpose of indicating the possession of a particular property or value but also, and in the first place, for the purpose of realizing a set of attributable properties or values.[3]

18.4.1 *A possible objection*

Suppose that $W \subset V'$, where V' is the spatial complement of V, and that the presence of O in V is indicated. Is not O's absence from W indicated as well? Are we not entitled to infer that the proposition "O is in W" has a truth value—namely, "false"?

Because regions of space do not exist by themselves, the answer is negative. If W is not realized by being the sensitive region of a detector in the broadest sense of the word—anything capable of indicating the presence of something somewhere—then W does not exist, and if it does not exist, then the proposition "O is in W" cannot be in possession of a truth value. Neither the property of being inside W nor the property of being outside W is available for attribution to O. All we can infer from O's indicated presence in V is the truth of a counterfactual: if W *were* the sensitive region of a detector D, then O would not be detected by D.

18.5 Spatiotemporal distinctions: Not all the way down

In a non-relativistic world, the exact localization of a particle implies an infinite momentum dispersion and, consequently, an infinite mean energy. In a relativistic world, the attempt to produce a strictly localized particle

---

[3]For instance, when measuring a spin component, the apparatus is needed not only to indicate the component's value but also to realize the axis with respect to which the component is defined.

results instead in the creation of particle-antiparticle pairs. It is therefore safe to say that no material object ever has a sharp position (relative to any other object). This implies something of paramount importance: the spatiotemporal differentiation of the physical world is incomplete—it does not go "all the way down."

To see what exactly this means, let $\mathbb{R}^3(O)$ be the set of (imaginary) exact positions relative to some object O. If no material object ever has a sharp position, we can conceive of a partition of $\mathbb{R}^3(O)$ into finite regions that are so small that none of them is the sensitive region of an actually existing detector. Hence we can conceive of a partition of $\mathbb{R}^3(O)$ into sufficiently small but finite regions R_k of which the following is true: there is no object Q and no region R_k such that the proposition "Q is inside R_k" has a truth value. In other words, there is no object Q and no region R_k such that R_k exists for Q. But if a region of space does not exist for any material object, it does not exist at all. The regions R_k—or the distinctions we make between them—correspond to nothing in the actual world. They exist solely in our minds.

What holds for the world's spatial differentiation also holds for its temporal differentiation. The times at which observables possess values, like the values themselves, must be indicated in order to exist. Clocks are needed not only to indicate time but also, and in the first place, to make times available for attribution to indicated values. Since clocks indicate times by the positions of their hands, the world's incomplete temporal differentiation follows from its incomplete spatial differentiation.[4]

What about the temporal differentiation of a quantum system? If the times that exist for it are the indicated times of possession of a property (by the system) or of a value (by an observable pertaining to the system), then the interval between two successive such times only exists as an undifferentiated whole. This is another reason why we need to reject the so-called eigenstate–eigenvalue link (Sec. 16.3). Not only b must be made attributable by some apparatus before it can be attributed to B, but also t must be made attributable by some clocklike apparatus before it can be the time at which b is possessed by B.

---

[4]Digital clocks indicate times by transitions from one reading to another, without hands. The uncertainty principle for energy and time, however, implies that such a transition cannot occur at an exact time, except in the unphysical limit of infinite mean energy [Hilgevoord (1998)].

18.6   The shapes of things

There is one notion that is decidedly at odds with the incomplete spatial differentiation of the physical world. It is the notion that the "ultimate constituents" of matter are pointlike (or, God help us, stringlike [Green *et al.* (1988)]). A fundamental particle—in the context of the Standard Model, a lepton or a quark—is a particle that lacks internal structure. It lacks internal relations; equivalently, it lacks parts. This could mean that it is a pointlike object, but it could also mean that it is *formless*.[5]

What does the theory have to say on this issue? It obviously favors the latter possibility, inasmuch as nothing in the formalism of quantum mechanics refers to the shape of an object that lacks internal structure.

And experiments? While they can provide evidence of internal structure, they cannot provide evidence of the absence of internal structure. Hence they cannot provide evidence of a pointlike form.

The notion that an object without internal structure has a pointlike form—or *any* form—is therefore unwarranted on both theoretical and experimental grounds. In addition, it explains nothing. Specifically, it does not explain why a composite object—be it a nucleon, a molecule, or a galaxy—has the shape that it does, inasmuch as all empirically accessible forms are fully accounted for by the relative positions (and orientations) of their material constituents. All it does is encumber our efforts to make sense of the quantum world with a type of form whose existence is completely unverifiable, which is explanatorily completely useless, and which differs radically from all empirically accessible forms. By rejecting this notion we obtain an appealingly uniform concept of form, since then *all* forms resolve themselves into sets of spatial relations—between parts whose forms are themselves sets of spatial relations, and ultimately between formless parts.

18.7   Space

Consider once more the fuzzy positions in Fig. 11.4. Does the expanse over which these positions are "smeared out" have parts? If it had, the positions themselves would have parts; they would be divided by the parts of space. But this makes no sense. One can divide an object, and thereby create as

---

[5]In order to leave a visible trace (a string of bubbles in a bubble chamber, a trail of droplets in a cloud chamber, or some such) an electron does not need a shape; it only needs to be *there*. In fact, it is was where it was only because its past whereabouts are indicated by bubbles or droplets or some such.

many positions as there are parts—or create as many relative positions as there are pairs of parts—but one cannot divide a position. The expanse over which a fuzzy position is probabilistically distributed therefore lacks parts. This confirms a conclusion we arrived at in Sec. 18.4: physical space cannot be something that has parts. If at all we think of it as an *expanse*, we must think of it as intrinsically undivided.

Alternatively we may attribute spatiality not to an expanse (to which relative positions owe their spatial character) but to the relative positions themselves, as a *quality*[6] they share. If we do so, we have no need to posit an independently existing expanse. Space itself can then be thought of as the set of all relations that share this quality, i.e., as the totality of spatial relations that exist between material objects.[7]

If we think this through, we arrive at the following conclusions: Space contains the forms of all things that have forms, for the totality of spatial relations contains—in the proper, set-theoretic sense of containment—the specific sets of spatial relations that constitute material forms. But it does not contain the corresponding relata—the formless "ultimate constituents" of matter. And if we give the name of "matter" to these "ultimate constituents," it does not contain matter.

---

[6]Like pink or turquoise, spatial extension is a qualitative property that can only be defined by ostentation—by drawing attention to something of which we are directly aware. While it can lend a phenomenal quality to numbers, it cannot be reduced to numbers. If you are not convinced, try to explain to my friend Andy, who lives in a spaceless world, what phenomenal space is like. Andy is good at math, so he understands you perfectly if you tell him that space is like the set of all triplets of real numbers. But if you believe that this gives him a sense of the expanse we call space, you are deluding yourself. *We* can imagine triplets of real numbers as points embedded in space; he can't. *We* can interpret the difference between two numbers as the distance between two points; he can't. At any rate, he can't associate with the word "distance" the phenomenal remoteness it conveys to us.

[7]This way of thinking is close to relationism, the doctrine that space and time are a family of spatial and temporal relations holding among the material constituents of the universe. A common objection to this doctrine is the claim that it fails to accommodate inertial effects. See Dieks (2001a,b) for a refutation of this claim.

Chapter 19

The macroworld

Making sense of the theoretical formalism of quantum mechanics calls for a judicious choice on our part. We need to identify that substructure of the theory's entire structure to which independent reality can be attributed. If observables have values only if (and only to the extent that) they are measured, this cannot be the microworld, nor any part thereof. The microworld is what it is because of what happens or is the case in the macroworld, rather than the other way round, as we are wont to think. This leaves us with the macroworld as the only structure to which independent reality can be attributed. It also leaves us with the task of defining the macroworld and rigorously distinguishing it from the microworld.

A definition, to begin with: by a *classically predictable position* I shall mean a position that can be predicted on the basis of (i) a classical law of motion and (ii) all relevant value-indicating events.

An observation, next: The possibility of obtaining evidence of the departure of an object O from its classically predictable position calls for detectors whose position probability distributions are narrower than O's— detectors that can probe the region over which O's fuzzy position extends. For objects with sufficiently sharp positions, such detectors do not exist. For the objects commonly and loosely referred to as "macroscopic," the probability of obtaining evidence of departures from their classically predictable motion will thus be low. Hence *among* these objects, there will be many of which the following is true: every one of their indicated positions is consistent with every prediction that can be made on the basis of previously indicated properties and a classical law of motion. These are the objects that truly deserve the label *macroscopic*. To permit a macroscopic object— e.g., the proverbial pointer needle—to indicate the value of an observable, one exception has to be made: its position may change unpredictably if and when it serves to indicate a property or a value.

We are now in position to define the *macroworld* unambiguously as the totality of relative positions between macroscopic objects. Let's shorten this to *macroscopic positions*. By definition, macroscopic positions never evince their fuzziness (in the only way they could, through departures from classically predicted values). Macroscopic objects therefore follow trajectories that are only *counterfactually* fuzzy: their positions are fuzzy only in relation to an imaginary background that is more differentiated spacewise than is the actual world. This is what makes it legitimate to attribute to the macroworld a reality independent of anything external to it—such as the consciousness of an observer[1]—and not merely "for all practical purposes" but strictly. And this in turn allows us to state the manner in which measurement outcomes are indicated. They are indicated by departures of macroscopic positions from their respective classical laws of motion.

But cannot the information provided by an outcome-indicating position be lost? A position that has departed from a classical law of motion once, to indicate a measurement outcome, may do so again, and may thereby lose the information about the outcome. This however does not mean that no record of the outcome persists. For the positions of macroscopic objects are abundantly monitored. Suppose that at the time t_2 a macroscopic position loses information about an outcome that it acquired at the time t_1. In the interim, a large number of macroscopic positions—among them macroscopic positions in the stricter sense—have acquired information about this position, and hence about the outcome that was indicated by it. Even if this position ceases to indicate the outcome, a record of the outcome persists.

I am not saying that macroscopic positions are exempted from our conclusion that to *be* is to be *measured*. Where macroscopic positions are concerned, this conclusion is not false but irrelevant. While even the Moon has a position only because of the myriad of "pointer positions" that betoken its whereabouts, macroscopic positions indicate each other's values so abundantly, so persistently, and so sharply that they are only counterfactually fuzzy. This is what makes it possible (and perfectly legitimate) to think of the positions of macroscopic objects as forming a self-contained

---

[1] In the context of quantum mechanics, consciousness has mainly been invoked to explain the so-called "collapse of the wave function" [Goswami (1995); Lockwood (1989); London and Bauer (1939); Squires (1990); Stapp (2001); von Neumann (1955); Wigner (1961)]. While this offers a gratuitous solution to a pseudo-problem arising from the mistaken belief that wave functions evolve, it obfuscate the real interpretational problems, such as demonstrating the legitimacy of attributing an independent reality to the macroworld.

system—the macroworld—and to attribute to this system a reality that depends on nothing external to it.

Let me emphasize, in conclusion, the crucial role played by the incomplete spatiotemporal differentiation of the physical world in defining the macroworld and in demonstrating the legitimacy of attributing to it an independent reality.

Chapter 20

Questions of substance

If the properties of the microworld are what they are because of what happens or is the case in the macroworld, rather than the other way around, then we cannot think of particles, atoms, and such as *constituents* of the macroworld. Then what is it that constitutes the macroworld? And what is a particle if it is not a constituent of the macroworld?

20.1 Particles

What we know about particles is what we can infer from correlations between "detector clicks." If we perform a series of position measurements, and if every position measurement yields exactly one outcome (i.e., each time exactly one detector clicks), then we are entitled to infer the existence of an entity O which persists through time, to think of the clicks given off by the detectors as indicating the successive positions of this entity, to think of the behavior of the detectors as position measurements, and to think of the detectors as detectors.

Things already get more complicated if each time exactly two detectors click. Are we entitled to infer from this the existence of two persistent entities?

20.2 Scattering experiment revisited

Let us return to the scattering experiment discussed in Sec. 14.2. This featured two alternatives:

- $N \to W$ and $S \to E$.
- $N \to E$ and $S \to W$.

Under the conditions stipulated by Rule A (Sec. 5.1), the probability with which the particles scatter at right angles is the sum of two terms:

$$p_\perp = |A(N{\to}W, S{\to}E)|^2 + |A(N{\to}E, S{\to}W)|^2.$$

This is consistent with the view that whenever the incoming particles were moving northward and southward, respectively, and the outgoing particles were moving eastward and westward, respectively, what really happened was either of the alternatives.

Under the conditions stipulated by Rule B, the probability with which the particles scatter at right angles is given by

$$p_\perp = |A(N{\to}W, S{\to}E) + A(N{\to}E, S{\to}W)|^2.$$

In this case, we concluded, what really happened cannot be either of the alternatives. In other words, there is no answer to the question "Which outgoing particle is identical with which incoming particle?" But a question that has no answer is a meaningless question.

Here as elsewhere, the challenge is to learn to think in ways that do not lead to meaningless questions. The question "Which is which?" arises because we assume that initially there are *two* things, one moving northward and one moving southward, that in the end there are *two* things, one moving eastward and one moving westward, and that each of these things remains identical with itself.

What if we assumed instead that initially there is *one* thing moving both northward and southward, and that in the end there is *one* thing moving both eastward and westward? Startling though this assumption may be, it has this advantage that the meaningless question "Which is which?" can no longer be asked. It is, moreover, the conclusion to which our interpretational strategy points. For this requires us to think in such a way that the distinction we make between the alternatives does not correspond to anything in the actual world. But it is precisely the idea that there are two particles *over and above* two sets of initial properties denoted by N and S and two sets of final properties denoted by E and W, that compels us to regard the two alternatives as objectively distinct.

20.3   How many constituents?

Returning to where we left off in Sec. 20.1, if each time two detectors click, and if the question "Which of the particles detected at t_1 is identical with which of the particles detected at t_2?" lacks an answer, then we are in the presence of a *single* entity with the property of being in *two* places whenever we check—not a system "made up" of two things but one thing with the property of being in *two* places every time a position measurement is made using an array of detectors.

Adopting the definitions of substance and property introduced by Aristotle,[1] we may say that a quantum system is always a single substance, while the number of its so-called "components" or "constituent parts" is just one of its measurable properties. Whereas in a non-relativistic context this property is constant, in a relativistic setting it can come out different every time it is measured. Hence if we permit ourselves to think of the physical universe as a quantum system and to ask about the number of *its* constituent substances, we find that there is just one. The rest is properties. Quantum mechanics lends unstinting support to the central idea of all truly monistic ontologies: ultimately there is only one substance.

20.4 An ancient conundrum

Imagine that in front of you there are two exactly similar objects. Because they are in different places, they are different objects. But is their being in different places the *sole* reason for their being different objects?

For centuries philosophers have debated this question. Those uninitiated into the mysteries of the quantum world are inclined to think that the difference between the two objects does *not* boil down to their being in different places; there has to be another difference. But what could that be?

It has been argued that the two objects, in addition to being in different places, are different substances. But how can one substance—by itself, intrinsically, irrespective of attributes—differ from another? How—come to think of it—can there be more than one substance (considered out of relation to attributes by which substances could be distinguished)?

To escape this quandary, some philosophers have postulated the property of being "this very object." According to them, two exactly similar objects in different places are different not only because they are in different places but also because one has the property of being "this very thing" while the other has the property of being "that very thing." Demonstrative

---

[1]According to Aristotle, a property is that in the world to which a logical or grammatical predicate can refer, whereas a substance is that in the world to which *only* a logical or grammatical subject can refer.

determiners like "this" and "that," however, distinguish things by pointing at them, and this is the same as distinguishing things by their positions. So this "solution" is but a reformulation of the problem.

Thanks to quantum mechanics, we know now why such attempts were bound to fail. Quantum mechanics does not permit us to interpose a multitude of distinct substances between the one ultimately existing substance and the multitude of existing ("possessed") positions or the multitude of existing ("possessed") bundles of properties. Individuality is strictly matter of properties. (The same was in fact true of the deterministically evolving fantasy world of classical physics. But since the determinism of that world made it possible in principle to keep track of the identities of objects, it did not prevent people from associating a distinct substance with each object, however redundantly.)

20.5   A fundamental particle by itself

Consider, finally, a fundamental particle "by itself," out of relation to anything else. What can we say about it? Apart from pointing out that it lacks a form, and that space (considered as the totality of spatial relations existing between material objects) does not contain it (Sec. 18.7), the plain and simple answer is: nothing. For the properties that are attributable to fundamental particles are either relational, like positions and momenta, or characteristic of interactions, like coupling parameters (charges), or they have objective significance independent of conventions only as dimensionless ratios, like mass ratios. They all involve more than one particle.

According to a philosophical principle known as the *identity of indiscernibles*, what appears to be two things A and B is actually one and the same thing just in case there is no difference between A and B. Although there is nothing so obvious that a philosopher cannot be found to deny it, this principle strikes me as self-evident. If true, it implies that all fundamental particles considered by themselves, out of relation to anything else, are identical in the strong sense of *numerical identity*.[2] Hence if we think of fundamental particles as the "ultimate constituents of matter," there is a clear sense in which the actual number of ultimate constituents is one.

---

[2]Numerical identity contrasts with qualitative identity or exact similarity. Examples of numerical identity are (i) the evening star and the morning star, (ii) Clark Kent and Superman.

Chapter 21

Manifestation

21.1 "Creation" in a nutshell

If all fundamental particles in existence—considered by themselves, apart from their relations—are identical in the strong sense of numerical identity (Sec. 20.5), we are in a position to account for the coming into being of both matter and space in a manner that is elegant and economical by any standard. All we still need is a name for the one substance that every fundamental particle intrinsically is. We shall call it Ultimate Reality and abbreviate it to UR, mindful of the fact that the prefix "ur-" carries the sense of "original."

Here goes: by *entering into spatial relations with itself*, UR creates both matter and space, for space is the totality of existing spatial relations, while matter is the corresponding apparent multitude of relata—*apparent* because the relations are *self*-relations.

21.2 The coming into being of form

If fundamental particles are formless (Sec. 18.6), we are also in a position to understand the coming into being of form.

Forms in the most general sense are sets of spatial relations in more or less stable configurations. They come into existence through aggregation—the formation of composite objects or bound states. Because they exist in multi-dimensional configuration spaces, as probability distributions, they cannot be visualized (or cannot be visualized *except as* multi-dimensional probability distributions).

The smallest structures that *can* be visualized consist of the mean relative positions of a molecule's constituent nuclei—the sticks of your

251

chemistry teacher's balls-and-sticks models of molecules. What makes these structures visualizable is the fact that the fuzziness of the relative positions of the nuclei (as measured by the standard deviations of the corresponding probability distributions) is generally small compared to the relative positions themselves (as given by their mean values).

From simple molecules, a well-known hierarchy of objects of increasing complexity and/or size then reaches up to the familiar forms of macroscopic experience.

21.3   Bottom-up or top-down?

For at least twenty-five centuries, theorists—from metaphysicians to natural philosophers to physicists and philosophers of science—have tried to explain the world from the bottom up, starting from an ultimate multiplicity and using concepts of composition and interaction as their basic explanatory tools. And still it does not strike us that the attempt to model reality from the bottom up—whether on the basis of an intrinsically and completely differentiated space or spacetime, out of locally instantiated physical properties, or by aggregation, out of a multitude of individual substances—is at odds with what quantum mechanics is trying to tell us: that reality is structured from the top down, by a self-differentiation of UR that does not bottom out.

The reason why it does not bottom out is that the distinctions we make—be they of a spatial or a substantial kind—are warranted by nothing but property-indicating events (Sec. 16.3), and these do not license an absolute and unlimited objectification of our distinctions. If we conceptually partition the physical world into smaller and smaller regions, we reach a point where our distinctions between the regions no longer correspond to anything in the physical world (Sec. 18.5), and if we go on dividing material objects, they lose their individuality by ceasing to be re-identifiable.

The idea that reality is structured from the top down is traditionally associated with the concept of *manifestation*: there is an Ultimate Reality or Pure Being, which manifests the world or manifests itself as the world (without thereby losing its essential unity). It does not get divided by the existence of space, for if space is an expanse, it is undifferentiated, and if space is the totality of existing spatial relations, the corresponding (ultimate) relata are numerically identical.

Adopting the top-down paradigm of manifestation, we identify the macroworld with the *manifested world*. The so-called microworld, extending as it were between UR and the macroworld, is *instrumental* in the manifestation of the macroworld. Quantum mechanics affords us a glimpse "behind" the manifested world at formless and numerically identical particles, non-visualizable atoms, and partly visualizable molecules, which, instead of being the world's constituent parts or structures, are instrumental in its manifestation.

The fact that the microworld is what it is because of what happens or is the case in the macroworld, rather than the other way round, now presents itself in a new light. There is a clear-cut difference between the manifested world and what is instrumental in its manifestation, and it makes good sense that what is instrumental in the world's manifestation can only be described—in fact, can only be *defined*—in terms of the finished product, the manifested world. The process of manifestation is a transition from numerical identity to effective multiplicity, a progressive differentiation of the undifferentiated. What lies "behind" the manifested world is, to varying degrees, indefinite and indistinguishable. But in order to describe— and even define—the indefinite and indistinguishable, we have to resort to probability distributions over events that are definite and distinguishable, and such events only exist in the manifested world.

21.4   Whence the quantum-mechanical correlation laws?

One thing seems certain: the attempt to *causally*[1] explain the quantum-mechanical correlation laws puts the cart before the horse. For it is the correlation laws themselves that tell us why causal explanations work to the extent they do. They work in the macroworld, inasmuch as the correlations between macroscopic positions evince no statistical variations. Because macroscopic positions are only *counterfactually* fuzzy (Chap. 19), their correlations are effectively deterministic, and deterministic correlations lend themselves to causal interpretations. What lies "behind" the macroworld, on the other hand, is out of bounds for the concept of causation—at least the "ordinary" kind, which links objects or events across space and/or time.

It is equally clear that a fundamental theory cannot be explained with the help of a "more fundamental" theory. We sometimes speak loosely of a

---

[1]Let alone mechanistically: "There are no 'wheels and gears' beneath this analysis of Nature" [Feynman (1985)].

theory as being more (or less) fundamental than another, but strictly speaking "fundamental" has no comparative. If quantum mechanics is indeed the fundamental theoretical framework of physics—and while there may be doubts, nobody has the slightest idea what an alternative framework consistent with the empirical data might look like—this kind of explanation is ruled out.

The explanatory vacuum left by any fundamental theory is an embarrassment to those who like to imagine themselves "potentially omniscient," i.e., capable in principle of knowing the furniture of the universe and the laws by which this is governed. Hence their attempts to reify mathematical symbols and equations, to interpret them as representing physical entities or describing physical processes. Such attempts have had some measure of success in classical physics, but attempts to reify the quantum-mechanical correlation laws are patently absurd. And when the classical correlation laws are recognized as simplifications of the quantum-mechanical ones in a particular asymptotic regime (the classical limit), it becomes clear that even their reification was never more than a sleight of hand (Sec. 10.7).

There is also the question about the origin of the physical laws and of their efficacy. Under the paradigm of manifestation this question becomes tractable to a certain extent. If we grant UR the power to enter into spatial relations with itself and thereby manifest a world, we should also grant it the power to subject the spatial relations to specific laws.

21.5 How are "spooky actions at a distance" possible?

This leaves us with two questions. First, why do the laws of physics have the particular form that they do? This question will be addressed in the next chapter. Second, how are these laws possible? This is not the question of "how Nature does it," which has already been disposed of: what is capable of manifesting a world is also capable of putting laws into effect. But the quantum-mechanical correlations—especially those between the outcomes of measurements performed in spacelike relation—have aspects that make them seem outright impossible.

Consider again the three-particle state (13.17). By measuring the x components or the y components of the spins of two particles, we can predict with certainty the x component of the spin of the third particle. By measuring the x component of one spin and the y component of another spin, we can predict with certainty the y component of the third spin. And by measuring the z component of one spin, we can predict with certainty the z components of the two other spins. How is this possible considering that

(1) these measurements do not reveal pre-existent values but create their outcomes,
(2) the correlations between the outcomes are independent of the distance between the particles,
(3) the correlations between the outcomes are independent of the relative times of the measurements.

We tend to believe, as Einstein (1948) did, that "things claim an existence independent of one another" just in case they "lie in different parts of space." We therefore readily believe that the three particles in this experiment lie in different parts of space. But what is space? If we think of it as an expanse, then quantum mechanics does not permit us to think of it as having parts (Sec. 18.7). Considered as an expanse, it offers no ground on which the three particles could claim an existence independent of one another. On the contrary, instead of separating things, space (qua expanse) unites them by its lack of intrinsic multiplicity. If, on the other hand, we look on space as the totality of spatial relations existing between material objects, the question of spatial parts does not arise. Nor does a relation offer a ground for claiming independence for its relata. Quite the contrary. Given that the ultimate constituents of matter are identical in the strong sense of numerical identity, every fundamental particle is UR, and every spatial relation is a self-relation. Seen in this light, how could things possibly "claim an existence independent of one another"? They can not.[2]

---

[2]A number of parapsychologists [e.g., Radin (2006)] have claimed that parapsychological phenomena can be explained in quantum-mechanical terms. Yet quantum mechanics explains neither how its correlations work nor how they are possible in the first place. It only tells us *that* they are possible (inasmuch as the observed correlations agree with the predicted ones). All there is to the alleged quantum-mechanical "explanation" of parapsychological phenomena is this: if the quantum-mechanical correlations are possible, then the correlations observed by parapsychologists cannot be dismissed offhand.

Chapter 22

Why the laws of physics are just so

22.1 The stability of matter

In Sec. 8.2 we described "ordinary" material objects as

- having spatial extent (they "occupy space"),
- being composed of a (large but) finite number of objects without spatial extent (particles that do not "occupy space"),
- being stable (they neither explode nor collapse as soon as they are created).

What does the existence of such objects entail? "Ordinary" objects occupy as much space as they do because atoms and molecules occupy as much space as *they* do. In Sec. 4.3 we considered a hydrogen atom and found that it must satisfy the following conditions: its internal relative position r and the corresponding momentum p must both be fuzzy, and the product $\Delta r \, \Delta p$ of the standard deviations of the respective radial components of r and p must have a positive lower limit. Since there is no reason why the internal relative position of the hydrogen atom should distinguish itself in some essential way from other relative positions, we concluded that the existence of "ordinary" objects requires that relative positions and their corresponding momenta be fuzzy, and that the product of the standard deviations of a relative position and its corresponding momentum should have a positive lower limit.

The stability of bulk matter containing a large number N of atoms also requires that the energy and the volume occupied by $2N$ atoms be twice the energy and the volume occupied by N atoms. If one assumes that the force between electrons and nuclei varies as $1/r^2$, this linear law has been shown to hold provided that the Pauli exclusion principle (Sec. 14.5) holds.

The original proof is due to Dyson and Lenard (1967/1968). A significant simplification of this proof was found by Lieb and Thirring (1975).

Problem 22.1. (∗) *The classical force between pointlike charges is inverse proportional to the square of the distance between them.*

Since only fermions obey the exclusion principle, the validity of this linear law also requires that the constituents of "ordinary" matter—electrons and nucleons or electrons and quarks—be fermions (Sec. 14.5). This in turn requires that the constituents of "ordinary" matter have a half-integral spin of at least 1/2 (Sec. 12.3). If electrons and nuclei were bosons, the volume occupied by $2N$ atoms would *decrease* like $-N^{7/5}$, and matter would collapse into a superdense state in which "the assembly of any two macroscopic objects would release energy comparable to that of an atomic bomb," Dyson and Lenard wrote.

The stability of bulk matter further requires the existence of independent upper bounds on the fine structure constant $\alpha = e^2/\hbar c$ and the product $Z\alpha$, where Z is the atomic number [Lieb (1976, 2005); Lieb and Seiringer (2009)]. This means, in particular, that the number of protons in a nucleus must have an upper limit.

22.2   Why quantum mechanics (summary)

The proper—mathematically rigorous and philosophically sound—way to define and quantify a fuzzy observable is to assign nontrivial probabilities to the possible outcomes of a measurement of this observable (Sec. 8.2).

The classical probability calculus cannot accommodate nontrivial probabilities (Sec. 8.1). The most straightforward way—in fact, the all but inevitable way—to make room for nontrivial probabilities is to upgrade from a 0-dimensional point $\mathcal{P}$ to a 1-dimensional line $\mathcal{L}$. Instead of representing a probability algorithm by a point in a phase space $\mathcal{S}$, we represent it by a 1-dimensional subspace of a vector space (more specifically, a Hilbert space) $\mathcal{V}$.[1] And instead of representing elementary tests by subsets of $\mathcal{S}$, we represent them by the subspaces of $\mathcal{V}$ (Sec. 8.3). Because there exists a one-to-one correspondence between subspaces and the linear operators

---

[1]In Secs. 5.7 and 8.4.1 we observed that the stability of free particles further requires that $\mathcal{V}$ be a complex vector space. (Not all definitions of "Hilbert space" require that it should contain complex vectors.)

that project into them, we adopted as our first postulate that *measurement outcomes are represented by the projectors of a vector space.*

Because the product of the respective standard deviations of a relative position and the corresponding momentum has a positive lower limit, it is impossible to simultaneously measure both quantities with arbitrary precision: the two quantities are *incompatible.* The formal definition of incompatibility, given in Sec. 8.5, led to our second postulate: *the outcomes of compatible elementary tests correspond to commuting projectors.*

In Sec. 8.6 we considered two measurements, one with three possible outcomes $\mathcal{A}$, $\mathcal{B}$, and $\mathcal{C}$, and one with two possible outcomes $\mathcal{A} \cup \mathcal{B}$ and $\mathcal{C}$. We thus had that $p(\mathcal{A}) + p(\mathcal{B}) + p(\mathcal{C}) = 1$ and that $p(\mathcal{A} \cup \mathcal{B}) + p(\mathcal{C}) = 1$. Obvious though it may seem, it does not follow that $p(\mathcal{C})$ in the former equation equals $p(\mathcal{C})$ in the latter. It was however possible to postulate the equality of the two probabilities, and so we did—no need to make the world stranger than it is. Hence our third postulate: *If $\hat{\mathbf{A}}$ and $\hat{\mathbf{B}}$ are orthogonal projectors, then the probability of the outcome represented by $\hat{\mathbf{A}} + \hat{\mathbf{B}}$ is the sum of the probabilities of the outcomes represented by $\hat{\mathbf{A}}$ and $\hat{\mathbf{B}}$, respectively.*

These three postulates are sufficient to prove the trace rule, according to which the probability of obtaining the outcome represented by the projector $\hat{\mathbf{P}}$ is given by $p(\hat{\mathbf{P}}) = \text{Tr}(\hat{\mathbf{W}}\hat{\mathbf{P}})$, where $\hat{\mathbf{W}}$ is a unique operator whose properties are listed in Secs. 8.8 and 8.10. The trace rule tells us (i) that the probabilities of the possible outcomes of measurements are encoded in a *density operator* $\hat{\mathbf{W}}$ and (ii) how they can be extracted from $\hat{\mathbf{W}}$. In Sec. 8.11 we established how the density operator is determined by actual measurement outcomes, and in Sec. 8.12 we learned how probabilities depend on the times of measurements.

In the last section of Chap. 8 we derived the two Rules that we postulated in the first section of Chap. 5. From there, Rule B led us to the particle propagator (5.10),

$$\langle \mathbf{r}_B, t_B | \mathbf{r}_A, t_A \rangle = \int \mathcal{D}\mathcal{C} \, Z[\mathcal{C} | \mathbf{r}_A, t_A \to \mathbf{r}_B, t_B],$$

where $Z[\mathcal{C} | \mathbf{r}_A, t_A \to \mathbf{r}_B, t_B]$ is some complex-valued functional of spacetime paths from $(\mathbf{r}_A, t_A)$ to $(\mathbf{r}_B, t_B)$. For a free and stable particle this takes the form (5.16),

$$Z[\mathcal{C} | \mathbf{r}_A, t_A \to \mathbf{r}_B, t_B] = e^{ibs[\mathcal{C} | \mathbf{r}_A, t_A \to \mathbf{r}_B, t_B]},$$

where the real-valued functional $s[\mathcal{C} | \mathbf{r}_A, t_A \to \mathbf{r}_B, t_B]$ is the length of $\mathcal{C}$ as defined by some spacetime geometry. To pin down this spacetime geometry, we then turned to the special theory of relativity.

22.3   Why special relativity (summary)

In Sec. 6.1 we learned why there is no such thing as an absolute position, an absolute orientation, an absolute time, or absolute rest. This—the content of the principle of relativity—took us straight to the Lorentz transformations in their general form (Sec. 6.2). In this form they contain a constant K which (if nonzero) has the dimension of an inverse velocity squared. Since the absolute value of K (if nonzero) depends on conventional units, we were left with exactly three physically distinct possibilities: $K > 0$, $K = 0$, and $K < 0$. The option $K > 0$ was ruled out in Sec. 6.4, and Sec. 6.8 made it clear that $K = 0$ can be valid only as an approximation to the special-relativistic form of the Lorentz equations, which is characterized by $K < 0$.

22.4   Why quantum mechanics (summary continued)

$K < 0$ gave us the wanted length functional (albeit measured in units of time):

$$s[\mathcal{C}] = \int_{\mathcal{C}} ds = \int_{\mathcal{C}} \sqrt{dt^2 - (dx^2 + dy^2 + dz^2)/c^2} = \int_{\mathcal{C}} dt\sqrt{1 - v^2/c^2}\,,$$

where $c = \sqrt{-1/K}$ is the speed of light (Sec. 6.6).

In Sec. 7.1 we learned that for a particle that is stable but not free, the amplitude associated with an infinitesimal path segment has the form

$$Z(t, \mathbf{r}, dt, d\mathbf{r}) = e^{(i/\hbar)\,dS(t,\mathbf{r},dt,d\mathbf{r})}.$$

(The division by $\hbar$ ensures that the action differential dS is measured in its conventional units.) dS is homogeneous (of first degree) in the differentials dt and $d\mathbf{r}$ (Eq. 7.7). The scope of possible effects on the motion of a particle is thereby limited to such modifications of the action differential associated with a free particle, $dS = -mc^2 ds$, as preserve both the homogeneity expressed by Eq. (7.7) and the invariance of dS as a 4-scalar (Sec. 7.2). The most straightforward such modification consists in the addition to $dS = -mc^2 ds$ of the scalar product of a 4-vector field $\vec{A} = (V, \mathbf{A})$ and the path element $d\vec{r} = (c\,dt, d\mathbf{r})$. How to get from here to the Schrödinger equation was spelled out in Secs. 7.2–7.4.

22.5   The classical or long-range forces

In Sec. 6.7 we followed Feynman's suggestion to associate a clock with every spacetime path contributing to the particle propagator (5.10). As the particle travels along such a path, the clock ticks. Although this happens in our imagination only, it reveals a deep connection between the quantum-mechanical probability calculus and the metric properties of the world. The rates at which particles tick not only lay the foundation for the world's metric properties but also make it possible to formulate ways of influencing the behavior of particles.

The world's metric properties are founded on the fact that the rates at which free particles tick in their rest frames—essentially, their masses—are constant. This makes it possible to introduce a global system of spacetime units. While there may be no global inertial frame, there will be local ones, and they will mesh as described by a pseudo-Riemannian spacetime geometry. Contrariwise, if the masses of the electrons (or protons, or neutrons) inside any given atom would depend on the spacetime paths along which they arrived at their current location, the Pauli exclusion principle would not hold, and an essential condition for the stability of matter would be missing [Marzke and Wheeler (1964)].

The only way to influence the probability of finding at one spacetime location a scalar particle last found at another location, is to modify the rate at which it ticks as it travels along the paths connecting the two locations. The number of ticks associated with an infinitesimal path segment defines a Finsler geometry $dS(dt, d\mathbf{r}, t, \mathbf{r})$ [Antonelli *et al.* (1993); Rund (1969)]. This can be influenced in two ways, and so, therefore, can the behavior of a scalar particle: a species-specific way represented by the 4-vector field A_i, which bends geodesics relative to local inertial frames, and a species-independent way represented by the metric g_{ik}, which bends the geodesics of the pseudo-Riemannian spacetime geometry: $dS = m\sqrt{g_{ik}dx^i dx^k} + qA_i dx^i$.

Since the positions of the sources of A_i and g_{ik} are fuzzy, the components of these fields cannot be sharp. We take this into account by summing over paths in the corresponding configuration spacetimes. This calls for the addition of terms that only contain the fields; these are determined by constraints spelled out in Secs. 10.2 and 10.4.

A renormalizable quantum theory of gravity does not (yet) exist. Nor is it clear why a quantum version of general relativity ought to be renormalizable, considering that such a theory would have an inbuilt cutoff. For if the metric becomes fuzzy, so do the distances between spacetime locations.

And if these become fuzzy, so do the spacetime locations themselves, inasmuch as they are defined by the distances between them (Sec. 10.4). On scales on which the fuzziness of the metric becomes significant, the very concept of "scale" appears to lose its meaning.

To arrive at QED, we still need the Dirac equation, and we need to take account of the fuzziness of particle numbers, due to the fact that only charges are conserved. To obtain the appropriate equation for a particle with spin, we need to let the wave function have more than one component. This equation will therefore be a matrix equation. The simplest version is linear in the operators of energy and momentum. If we require in addition that each component of ψ satisfy the Klein–Gordon equation, we find that the lowest possible number of components is four, and that this number yields the appropriate equation for a spin-1/2 fermion—the Dirac equation (15.5). How the fuzziness of particle numbers can be taken into account is described in Sec. 15.6.

The stability of matter rests on the stability of atoms, and without the electromagnetic force atoms would not exist. The stability of matter further rests on the validity of quantum mechanics, and quantum mechanics presupposes not only value-indicating events but also persistent records of such events (Chap. 19). The existence of such events and such records does not seem possible without the relatively hospitable environment of a planet, and without gravity planets would not exist.

If we further take into account the evolutionary nature of our world, we have reason to expect that the requisite variety of chemical elements is not present from the start, and we can point to the fact that without gravity there would be no stars to synthesize elements beyond beryllium.[2]

22.6 The nuclear or short-range forces

Quantum mechanics presupposes measurement outcomes. It thus presupposes macroscopic objects, including objects that can perform the function of a measurement apparatus. Yet it seems all but certain that the existence of outcome-indicating devices requires a variety of chemical elements well beyond the four—hydrogen, helium, and a sprinkling of lithium and beryllium—that are known to have existed before the formation of stars. If so,

---

[2] "Evolutionary" is used here in a sense different from that in which the time-dependence of a quantum state is frequently misconstrued. At a minimum, an evolutionary world has a beginning and grows in complexity.

QED, which is based on the gauge group U(1), and gravity are both necessary parts of the laws governing UR's spatial self-relations, but they are not sufficient. Interactions of a different kind are needed to carry further the nucleosynthesis that took place before stars were formed. (Nucleosynthesis is the process of creating atomic nuclei from protons and neutrons.)

By hindsight we know that QCD (Sec. 15.9.4), based on the gauge group SU(3), bears much of the responsibility for nucleosynthesis, both primordial and inside stars. (The elements created in stellar nucleosynthesis range in atomic numbers from six to at least ninety-eight, i.e., from carbon to at least californium.) QCD is further responsible for the formation of the first protons and neutrons, which are thought to have condensed from the quark-gluon plasma that emerged from the Big Bang—the hot and dense condition from which the Universe began to expand some 13–14 billion years ago.

But if most of the chemical elements are created inside stars, how do they get out to form planets? Sufficiently massive stars end their lives with an explosion, as Type II supernovae, spewing the elements created in their interiors into the interstellar medium—dust clouds that have been produced by explosions of earlier generations of stars. New stars and eventually planets condense from these clouds, sometimes triggered by shock waves generated by supernova explosions. It took many stellar life cycles to build up the variety and concentration of heavy elements that is found on Earth.

A Type II supernova occurs when the nuclear fusion reactions inside a star are depleted to the point that they can no longer sustain the pressure required to support the star against gravity. During the ensuing collapse, electrons and protons are converted into neutrons and neutrinos. The central core ends up either a neutron star or a black hole, while almost all of the energy released by its collapse is carried away by prodigious quantities of neutrinos, which blow off the star's outer mantle. But if neutrinos are crucial for the release into the interstellar medium of the products of stellar nucleosynthesis, then so is the weak force, which is based on the gauge group SU(2).

Supernova explosions not only release the products of stellar nucleosynthesis but themselves contribute significantly to the synthesis of the heavier elements. The weak force, for its part, not only is crucial to supernova explosions but also plays an essential part in stellar nucleosynthesis.

It is also clear why the "carriers" of the weak force need to have large masses, and why a mathematical procedure like the Higgs mechanism

(15.9.6) is needed to "create" them. The range of a force is characteristically given by the Compton wavelengths of the particles mediating it. For a particle of mass m this equals h/mc. If the masses of the $W^{\pm}$ and the Z^0 were too small, the weak force would cause the beta decay of neutrons in the atomic nucleus through interactions with atomic electrons, as well as the decay of atomic electrons into neutrinos through interactions with nucleonic quarks. All matter would be unstable.

It thus appears safe to say that the well-tested physical theories are preconditions of the possibility of objects that (i) have spatial extent, (ii) are composed of a finite number of objects without spatial extent, and (iii) neither explode nor collapse as soon as they are created (Secs. 8.2 and 22.1). In other words, the existence of such objects appears to require quantum mechanics, special relativity, general relativity, and the standard model of particle physics, at least as effective theories.[3]

22.7 Fine tuning

The standard model of particle physics has about twenty-six freely adjustable parameters. Certain combinations of these parameters appear to be remarkably fine-tuned for life [Barrow and Tipler (1986); Gribbin and Rees (1989); Davies (2007)]. As Stephen Hawking (1988) wrote, "The laws of science, as we know them at present, contain many fundamental numbers.... The remarkable fact is that the values of these numbers seem to have been very finely adjusted to make possible the development of life."

Example 1: Supernova explosions only occur if a certain numerical relation involving the dimensionless coupling constants of the weak and gravitational interactions (α_w and α_g) and the proton-electron mass ratio m_p/m_e are approximately satisfied. If α_w were too large, the neutrinos released by the collapse of the core of a star could not reach the stellar envelope before losing most of their energy. And if α_w were too small, the neutrinos would escape with most of their energy. In either case the star would fail to explode.

Example 2: Stars rely on two mechanisms for transporting energy from their cores to their surfaces: radiation and convection. Astronomical observations indicate that only stars which are at least partially convective

---

[3] An effective theory includes appropriate degrees of freedom to describe physical phenomena occurring above a given length scale, while ignoring substructure and degrees of freedom that exist or may exist at shorter distances.

have planets. On the other hand, only radiative stars explode. So stars of both types are needed, and this calls for a delicate balance between the dimensionless coupling constants of electromagnetism and gravity and the proton-electron mass ratio. If α_g were slightly greater, only radiative stars would exist, and if α_g were slightly smaller, only convective stars would exist.

Example 3: The synthesis of carbon is a two-step process. The first step is the formation of a ^{8}Be nucleus out of two ^{4}He nuclei (alpha particles), the second the formation of a ^{12}C nucleus out of the ^{8}Be nucleus and another ^{4}He nucleus. The probability of this process would be extremely small, were it not for two "coincidences": the ^{8}Be ground state has almost exactly the energy of two alpha particles, and ^{8}Be $+\,^4$He has almost exactly the energy of an excited state of ^{12}C. In other words, the ^{8}Be ground state "resonates" with a system comprising two alpha particles, and the excited ^{12}C state "resonates" with a systems comprising ^{8}Be and ^{4}He. The existence of the second resonance was predicted by Fred Hoyle before its actual observation, based on the observed abundance of carbon in the Universe and the necessity for it to be formed in stars. The energy at which this resonance occurs depends sensitively on the interplay between the strong and the weak nuclear interactions. If the strong force were slightly stronger or slightly weaker—by just 1% in either direction—then the binding energies of the nuclei would be different, and the requisite resonance would not exist. In that case, there would be no carbon or any heavier elements anywhere in the Universe. "I do not believe," Hoyle (1959) concluded, "that any scientist who examined the evidence would fail to draw the inference that the laws of nuclear physics have been deliberately designed with regard to the consequences they produce inside the stars."

It is self-evident that the features of the Universe impose constraints on its laws. If, for example, life has evolved, and if general relativity and the theories included in the Standard Model are valid, then the adjustable parameters of these theories must be so constrained as to allow for the evolution of life. This truism is one version of the (weak) *anthropic principle*. What is nevertheless remarkable is the number of constraints that have been uncovered and, in consequence, the extent to which these adjustable parameters are fine-tuned for the existence of life in the Universe.

If, on the other hand, there are objects that (i) have spatial extent, (ii) are composed of a finite number of objects without spatial extent, and (iii) neither explode nor collapse as soon as they are created, then we must have quantum mechanics, special relativity, general relativity, and the

Standard Model—the latter at least as effective theories—and the adjustable parameters of these theories must be so constrained as to allow for the existence of such objects. Quantum mechanics presupposes macroscopic objects, including objects that can function as outcome-indicating devices, and it seems all but certain that the existence of such devices calls for elements whose existence depends on stellar nucleosynthesis and supernova explosions. In other words, it calls for some of the same fine tuning that has been shown to be necessary for life.

Chapter 23

Quanta and Vedanta

Why are objects that have spatial extent composed of finite numbers of objects without spatial extent? And if there is any merit to the one-sentence creation story told in Sec. 21.1, how does UR enter into spatial relations with itself? As we shall see, these questions are closely related.

Science operates within a metaphysical framework that formulates questions and interprets answers. This metaphysical framework is not testable by the methods of science. Where quantum mechanics is concerned, the most that can be said is that it makes more sense when placed in the right (or an adequate) framework than when placed in the wrong (or an inadequate) framework. The following appraisals appear to be clear indications of the use of an inadequate metaphysical framework:

> I think it is safe to say that no one understands quantum mechanics. [Feynman (1967)]
>
> [Quantum theory] makes absolutely no sense. [Penrose (1986)]
>
> It is often stated that of all the theories proposed in this century, the silliest is quantum theory. [Kaku (1995)]

Whereas the top-down framework put forward here (Sec. 21.3) takes quantum theory's *verifiable* probability assignments as its point of departure, the standard, bottom-up approach leads to interpretations that contain assumptions about what happens between measurements—assumptions that by definition cannot be put to the test. The inadequacy of this approach was demonstrated in Chaps. 16 and 17.

A top-down framework has other advantages: it can make better sense of the reality of consciousness, it may be the only framework that can adequately deal with the reality of quality and value, and it affords deeper insights into the nature of evolution.

23.1 The central affirmation

For the remainder of this text I shall take my cue from a more than millennium-long philosophic tradition known as Vedanta, which is founded on a group of Indian scriptures, the Upanishads [Phillips (1995); Sri Aurobindo (2001, 2003, 2005)]. The central affirmation of this tradition is that there is an Ultimate Reality, and that this relates to the world in a threefold manner: it is the substance that constitutes the world, it is a consciousness that contains the wolrd in its totality, and it is (subjective speaking) an infinite delight or bliss and (objectively speaking) an infinite quality or value that expresses and experiences itself in the world.

Two important observations can already be made at this point. Within a bottom-up framework of thought, what ultimately exists is a multitude of entities (atoms, fundamental particles, spacetime points, you name them) without intrinsic quality or value. In many traditions this multiplicity is fittingly referred to as "dust." In such a framework it is obviously hard to give a non-reductive account of quality and value. In a top-down framework of the Vedantic kind, on the other hand, quality and value have their roots in the very heart of reality.

The second observation is that the substance that constitutes the world and the consciousness that contains the world are one. UR is one, but the world exists both *by* it and *for* it. When we think of the world as existing by it, we think of it as a substance; when we think of the world as existing for it, we think of it as a consciousness. This identity does not mean that the consciousness we are familiar with is identical with any material structure or function. What it means is that the consciousness we are familiar with has its roots in that other consciousness. If this is one with the substance that constitutes the world—which, remember, is what every fundamental particle intrinsically is—then we have a fighting chance of understanding how anything material could be conscious (that is, how it could possess the private, first-person, subjective aspect of consciousness). Otherwise we don't.[1]

---

[1] "Nobody has the slightest idea how anything material could be conscious. Nobody even knows what it would be like to have the slightest idea about how anything material could be conscious. So much for the philosophy of consciousness" [Fodor (1992)]. Nothing much has changed since this was written.

23.2 The poises of creative consciousness

There exist a series of poises of relation between the consciousness that contains the world and the world contained in it.

The original poise features a single conscious self, which is coextensive with the world. Since in this poise the subject is wherever its objects are, no distances exist between the seer and the seen. There is an expanse of some kind—otherwise there would be no world—but this has neither the quality of space nor that of time.

In a secondary poise, consciousness bifurcates: the self distantiates itself from the content. This allows consciousness to apprehend its content from a location within its content, perspectively, and to do this many times, thereby taking on the aspect of a multitude of localized selves. It is here, in this poise, that the three dimensions of space—viewer-centered depth and lateral extent—come into being, for objects are no longer seen from within, by identity with the all-constituting substance, but from outside, as presenting their surfaces.

It is here that consciousness becomes distinct from substance. For whereas in the primary poise the world's properties exist indistinguishably as determinations of a single substance and as content of a single consciousness, the properties of a conscious individual exist distinguishably as determinations of this particular individual (qua substance) and as content of many another individual (qua consciousness).

A third poise arises if the multiple concentration of consciousness, which has created the multitude of conscious selves, becomes exclusive. We all know first-hand a state of exclusive concentration, in which awareness is focused on a single object or task, while other goings-on are registered, or other tasks attended to, subconsciously, if at all. It is by a similar—albeit not as easily reversible—concentration that consciousness loses sight, in each individual self, of its identity with the other selves and with the single self of the primary poise.

Various degrees of exclusiveness are possible. A characterization of the main degrees can be obtained by thinking of the creative process—the transition from infinite quality to revealing form—as involving a couple of intermediate stages:

Infinite Quality → Expressive Idea → Executive Force → Revealing Form

When consciousness loses sight, in the individual self, of its identity with the single self of the primary poise, it also loses sight of its oneness with the infinite quality/delight at the heart of existence. It then becomes the center of a quantitative and finite action that is no longer aware of its original purpose, which is to develop infinite quality into expressive ideas. Whatever infinite quality is still expressed, is now received subliminally, unbeknown to the individual.

If, by a further deepening of the concentration of consciousness, this action is also excluded, we arrive at a world whose individuals execute expressive ideas unconsciously.[2] And if the exclusive concentration is carried to its ultimate extreme, then even the executive force that was active in the individual falls dormant. Because this is instrumental in creating and maintaining individual forms, what remains is a multitude of formless individuals. The stage for UR's adventure of evolution has been set. Welcome to the physical world!

This, then, could be the reason why objects that have spatial extent are composed of (finite numbers) of objects without spatial extent.

In a sense, evolution is the reverse of the sequence of essentially psychologically processes by which UR enters into spatial relations with itself and ends up a discrete multitude of formless relata. Life (individuals capable of executing ideas without being conscious of them) evolves first, then mind (conscious individuals unaware of the single self and the infinite quality at the heart of existence), and so will the higher poises of relation between consciousness and the world, eventually. There is however this difference: the evolution of life does not transform formless entities back into individuals capable of executing expressive ideas; instead it proceeds by aggregation, manifesting forms as sets of spatial relations between formless entities, and manifesting qualities with the help of forms. More generally, it proceeds by an ascent to a higher poise of relation and a partial but increasingly comprehensive integration of the constituents of the lower poise [Sri Aurobindo (2005), Bk. 2, Chap. 18].

---

[2]Think of the angiosperms (flowering plants) as examples, and don't let yourself be bamboozled into thinking that the beauty of a flower is but a device that serves to ensure the survival of a species. While in a bottom-up framework of thought, it is natural to end up by saying that qualities are "nothing but" quantities, in a top-down framework that has infinite quality at its core, quantities are "nothing but" means of manifesting qualities.

Appendix A

Solutions to selected problems

Problem 1.2 (Sally Clark). The fatally wrong assumption was that two sudden infant deaths in the same family are independent events. What the "expert" ignored was that there are factors—genetic, environmental, etc.— that may increase the probability of a sudden infant death. The first sudden infant death signals that such a factor is probably present. The occurrence of a second sudden infant death in the same family therefore has a higher probability, so the product rule for independent events cannot be applied.

Problem 1.3 (Monty Hall). Let D be the door originally chosen by the player, let H be the door opened by the host (revealing a goat), and let R be the remaining door. Before the host opens H, the probability that the Grand Price is behind D is $1/3$, and the probability $p(\overline{D})$ that the Grand Price is *not* behind D is $2/3$. Neither probability is affected by the host's opening H. What is affected is $p(H)$, which drops from $1/3$ to 0, and $p(R)$, which jumps from $1/3$ to $2/3$, now equaling $p(\overline{D})$. Accepting the host's offer to switch doors therefore *doubles* the player's chances to win the Grand Price.

Problem 1.4. To solve the previous problem, we did not have to invoke frequencies in order to conclude that switching doors doubles the probability of winning. Hence the correct answer is (i): winning after switching doors happens more often because it is more likely.

Problem 1.5. The answer depends on the rarity of the disease. Suppose that one person in 10,000 has it. This means that if a million people are tested, most probably there will be 99 true positives (99% of 100 people) and one false negative (1% of 100), and there will be 989,901 true negatives (99% of 999,900 people) and 9,999 false positives (1% of 999,900). The probability that a randomly picked person who tests positive actually has the disease is therefore less than 1%: the number of true positives divided by the number of all positives: $99/(9999 + 99) \approx 0.0098$.

Problem 1.6. The expected value is

$$\sum_{k=2}^{12} p(k)\, k = \frac{1}{36} \times 2 + \frac{1}{18} \times 3 + \frac{1}{12} \times 4 + \frac{1}{9} \times 5 + \frac{5}{36} \times 6 + \frac{1}{6} \times 7$$

$$+ \frac{5}{36} \times 8 + \frac{1}{9} \times 9 + \frac{1}{12} \times 10 + \frac{1}{18} \times 11 + \frac{1}{36} \times 12 = 7.$$

The squared deviations from the mean are $25, 16, 9, 4, 1, 0, 1, 4, 9, 16, 25$, respectively. Their mean is

$$\frac{1}{36} \times 25 + \frac{1}{18} \times 16 + \frac{1}{12} \times 9 + \frac{1}{9} \times 4 + \frac{5}{36} \times 1 + \frac{1}{6} \times 0$$

$$+ \frac{5}{36} \times 1 + \frac{1}{9} \times 4 + \frac{1}{12} \times 9 + \frac{1}{18} \times 16 + \frac{1}{36} \times 25 = \frac{35}{6},$$

and the root of this is $\sqrt{35/6} = 2.41523$.

Problem 3.1. Hint: imagine (i) three vectors $\mathbf{a}_x, \mathbf{a}_y, \mathbf{a}_z$ parallel to the x, y, and z axis, respectively, such that $\mathbf{a}_x + \mathbf{a}_y + \mathbf{a}_z = \mathbf{a}$ and (ii) another three vectors $\mathbf{b}_x, \mathbf{b}_y, \mathbf{b}_z$ parallel to the x, y, and z axis, respectively, such that $\mathbf{b}_x + \mathbf{b}_y + \mathbf{b}_z = \mathbf{b}$.

Problem 3.2. Since $\mathbf{a} \cdot \mathbf{b}$ is a scalar, we can choose a convenient coordinate system to calculate it. In a coordinate system in which $\mathbf{a} = (a, 0, 0)$ we have $\mathbf{a} \cdot \mathbf{b} = ab_x$ with $b_x = b \cos \theta$.

Problem 3.4. Hint: calculate the scalar product of $\mathbf{a} \times \mathbf{b}$ with $\mathbf{a}$ and $\mathbf{b}$.

Problem 3.6. Hint: use a coordinate system in which $\mathbf{a} = (a, 0, 0)$ and $\mathbf{b} = (b \cos \theta, b \sin \theta, 0)$.

Problem 3.8.
$$\lim_{\Delta x \to 0} \frac{[2(x + \Delta x)^2 - 3(x + \Delta x) + 4] - [2x^2 - 3x + 4]}{\Delta x} = 2x - 3.$$

Problem 3.9. Where $f''(x) > 0$, the slope of $f(x)$ increases, and the graph of $f(x)$ curves upwards. Where $f''(x) < 0$, the slope of $f(x)$ decreases, and the graph of $f(x)$ curves downwards. If at the same time $f'(x) = 0$, we have either a local maximum (if $f''(x) < 0$) or a local minimum (if $f''(x) > 0$).

Problem 3.11. Hint: apply the product rule (3.13) twice.

Problem 3.13. Hint: use the product rule (3.13) to calculate $(x^n x^{-n})'$.

Problem 3.14. Hint: use the product rule to calculate the derivative of $(x^{1/m})^m$.

Problem 3.18.

$$\exp(a+b) = 1 + (a+b) + \frac{(a+b)^2}{2!} + \frac{(a+b)^3}{3!} + \cdots$$

$$\exp(a)\exp(a) = \left[1 + a + \frac{a^2}{2!} + \frac{a^3}{3!} + \cdots\right]\left[1 + b + \frac{b^2}{2!} + \frac{b^3}{3!} + \cdots\right]$$

$$= (1) + (a+b) + \left(\frac{b^2}{2!} + ab + \frac{a^2}{2!}\right) + \left(\frac{b^3}{3!} + \frac{ab^2}{2!} + \frac{a^2 b}{2!} + \frac{a^3}{3!}\right) + \cdots$$

In the last line we have grouped terms of the same order and placed them in ascending order. (The order of a term is the sum of the powers of a and b.)

Problem 3.19.

$$e = \sum_{n=0}^{\infty} \frac{1}{n!} = 1 + 1 + \frac{1}{2} + \frac{1}{6} + \cdots = 2.718281828459\ldots$$

Problem 3.22. Hint: differentiate $e^{\ln f(x)}$.

Problem 3.23. Hint: wherever $\cos(x)$ is positive, its slope decreases as x increases (that is, its graph curves downward), and wherever $\cos(x)$ is negative, its slope increases as x increases (that is, its graph curves upward).

Problem 3.26. The antiderivative of x^2 is $x^3/3$. The value of the integral is therefore $2^3/3 - 1^3/3 = 7/3$.

Problem 3.27. The antiderivative of $1/r^2$ is $-1/r$. The value of the integral is therefore $-1/\infty + 1/r = 1/r$.

Problem 3.31. Hint: recall the Taylor series for $\cos(x)$ and $\sin(x)$.

Problem 3.32. Real part: $\cos(\pi/4) = 1/\sqrt{2}$; imaginary part: $\sin(\pi/4) = 1/\sqrt{2}$.

Problem 3.33. The first equation is equivalent to $x^2 = e^{i\pi/2}$. We need to find all angles that, when doubled, equal $\pi/2$. These are $\pi/4$ and $\pi/4 + \pi$. The second equation is equivalent to $x^3 = e^{i\pi}$. We need to find all angles that, when tripled, equal π. These are $\pi/3$, $\pi/3 + (2\pi)/3$, and $\pi/3 + (4\pi)/3$.

Problem 3.35. $e^{i\pi} + 1 = 0$.

Problem 5.1. Hint: use $|z|^2 = z\,z^*$ and remember Problem 3.34.

Problem 6.3. Hint: make use of Eq. (6.32).

Problem 6.7. Use $\tan\alpha = \sin\alpha/\cos\alpha$, multiply numerators and denominators by $\cos\alpha$, and remember that $\cos^2\alpha + \sin^2\alpha = 1$.

Problem 6.9. $dx = c\,dt$ implies $dx' = c\,dt'$.

Problem 8.7. Hint: make use of the orthonormality relations (8.9).

Problem 8.23. Hint: show with the help of the spectral decomposition that $\alpha_i^2 = \alpha_i$.

Problem 8.25. Hint: show with the help of Eq. (8.33) that $\alpha_k^2 < \alpha_k$.

Problem 8.26. Hint: $\mathrm{Tr}(\hat{\mathbf{W}}) = 1$.

Problem 8.28. Hint: make use of Eq. (8.40).

Problem 8.29. Owing to the invariance of the scalar product, we have that $\langle \hat{\mathbf{U}}v|\hat{\mathbf{U}}v \rangle = \langle v|v \rangle$. If $|v\rangle$ is an eigenvector of $\hat{\mathbf{U}}$ with eigenvalue λ, then $\lambda^*\lambda\langle v|v \rangle = \langle v|v \rangle$.

Problem 8.30. $|u\rangle = \hat{\mathbf{U}}^{\dagger}\hat{\mathbf{U}}|u\rangle = \hat{\mathbf{U}}^{\dagger}\lambda|u\rangle$ implies $\hat{\mathbf{U}}^{\dagger}|u\rangle = \lambda^{-1}|u\rangle$.

Problem 8.31. $0 = \langle u_1|\hat{\mathbf{U}}u_2 \rangle - \langle \hat{\mathbf{U}}^{\dagger}u_1|u_2 \rangle = (\lambda_2 - \lambda_1)\langle u_1|u_2 \rangle$.

Problem 10.1. Hint: the integral $\int_{\mathcal{C}} df$ only depends on the values of f at the endpoints of $\mathcal{C}$ (Problem 9.4).

Problem 10.2. Hint: $df = (\partial f/\partial t)\,dt + (\partial f/\partial\mathbf{r}) \cdot d\mathbf{r}$.

Problem 10.4.

$$\frac{dS}{dt} = -mc^2\sqrt{1 - \frac{\mathbf{v}^2}{c^2}} - qV + \frac{q}{c}\mathbf{A} \cdot \mathbf{v}$$

Problem 10.8. Hint: draw a spacetime diagram.

Problem 10.9. Hint: remember the antisymmetry of F_{ik}.

Problem 11.3. Hint: calculate $[\hat{x}, \hat{p}]\,\psi(x)$.

Problem 11.5. Hint: consider the action of these operators on a wave function.

Problem 11.6. Hint: make use of Problem 11.4.

Problem 12.1. Hint: apply the identity operator $|z_+\rangle\langle z_+| + |z_-\rangle\langle z_-|$ to $|z''_+\rangle$.

Problem 12.4. $\langle i|\hat{\mathbf{A}}^\dagger|k\rangle = \langle \hat{\mathbf{A}}i|k\rangle = \langle k|\hat{\mathbf{A}}|i\rangle^* = A_{ki}^*$.

Problem 12.5. Hint: the respective matrix representations of $|1\rangle$ and $|2\rangle$ are

$$\begin{pmatrix}1\\0\end{pmatrix} \quad \text{and} \quad \begin{pmatrix}0\\1\end{pmatrix}.$$

Problem 12.6.

$$\begin{pmatrix}t'\\x'\end{pmatrix} = \begin{pmatrix}\cos\alpha & \sin\alpha\\-\sin\alpha & \cos\alpha\end{pmatrix}\begin{pmatrix}t\\x\end{pmatrix}.$$

Problem 12.7.

$$\langle i|\hat{\mathbf{U}}^\dagger\hat{\mathbf{U}}|k\rangle = \sum_{j=1}^{N}\langle i|\hat{\mathbf{U}}^\dagger|j\rangle\langle j|\hat{\mathbf{U}}|k\rangle = \langle i|k\rangle.$$

Problem 12.16. Hint: make use of Eqs. (12.28) and (12.29).

Problem 12.17. There are four possibilities whose total probability is $|\langle z_+|a\rangle|^2$ and four possibilities whose total probability is $|\langle z_-|a\rangle|^2$. But

$$|\langle z_+|a\rangle|^2 + |\langle z_-|a\rangle|^2 = \langle a|z_+\rangle\langle z_+|a\rangle + \langle a|z_-\rangle\langle z_-|a\rangle = \langle a|a\rangle = 1.$$

Problems 13.6 and 13.7. Hint: make use of Eqs. (12.28–12.30).

Problem 13.13.

$$\hat{\mathbf{A}} = \begin{pmatrix}1 & 0\\0 & -1\end{pmatrix}, \quad \hat{\mathbf{B}} = \begin{pmatrix}1/5 & 2\sqrt{6}/5\\2\sqrt{6}/5 & -1/5\end{pmatrix}.$$

Problem 14.2. *BE dice*: The probability of a first 6 is 1/6. The probability of subsequently tossing another 6 is twice that of subsequently tossing a different number. The (conditional) probability of a second 6 therefore equals 2/7. Hence $p_{BE}(12) = (2/7)(1/6) = 1/21$. In all there are six ways of tossing equal numbers, each with a probability of 1/21, and there are fifteen ways of tossing different numbers, whose probabilities are given by

$(1 - 6/21)/15 = 1/21$. Accordingly, $p_{BE}(11) = 1/21$, $p_{BE}(10) = p_{BE}(9) = 2/21$, $p_{BE}(8) = p_{BE}(7) = p_{BE}(6) = 3/21$, $p_{BE}(5) = p_{BE}(4) = 2/21$, and $p_{BE}(3) = p_{BE}(2) = 1/21$.

FD dice: In all there are fifteen ways of tossing different numbers, each with probability $1/15$. Accordingly, $p_{FD}(12) = 0$, $p_{FD}(11) = p_{FD}(10) = 1/15$, $p_{FD}(9) = p_{FD}(8) = 2/15$, $p_{FD}(7) = 3/15$, $p_{FD}(6) = p_{FD}(5) = 2/15$, $p_{FD}(4) = p_{FD}(3) = 1/15$, and $p_{FD}(2) = 0$.

Problem 22.1. Hint: Use $(\partial/\partial \mathbf{r}) \cdot \mathbf{E} = 4\pi\rho$ (Eq. 10.20) and Gauss' law (Eq. 9.37).

Bibliography

Aharonov, Y. Bergmann, P.G. and Lebowitz, J.L. (1964). *Physical Review B* **134**, pp. 1410–1416; reprinted in Wheeler and Zurek (1983), pp. 680–686.

Aharonov, Y. and Bohm, D. (1959). *Physical Review* **115**, pp. 485–491.

Albert, D.Z. (1992). *Quantum Mechanics and Experience* (Harvard University Press), Chap. 7.

Antonelli, P.L., Ingarden, R.S. and Matsumoto, M. (1993). *The Theory of Sprays and Finsler Spaces with Applications in Physics and Biology* (Kluwer).

Arndt, M., Nairz, O., Vos–Andreae, J., Keller, C., van der Zouw, G. and Zeilinger, A. (1999). *Nature* **401**, pp. 680–682.

Aspect, A. (2002). In R.A. Bertlmann and A. Zeilinger, *Quantum [Un]speakables —From Bell to Quantum Information* (Springer).

Barrow, J. and Tipler, F. (1986). *The Anthropic Cosmological Principle* (Oxford University Press).

Bell, J.S. (1964). *Physics* **1**, pp. 195–200.

Bell, J.S. (1966). *Reviews of Modern Physics* **38**, pp. 447–452; reprinted in Wheeler and Zurek (1983), pp. 397–402, and in Bell (1987), pp. 1–13.

Bell, J.S. (1987). *Speakable and Unspeakable in Quantum Mechanics* (Cambridge University Press).

Bell, J.S. (1990). In A.I. Miller (ed.), *62 Years of Uncertainty* (Plenum), pp. 17–31.

Bohm, D. (1951). *Quantum Theory* (Prentice Hall).

Bohm, D. (1952). *Physical Review* **85**, pp. 166–193.

Bouwmeester, D., Pan, J–W., Daniell, M., Weinfurter, H. and Zeilinger, A. (1999). *Physical Review Letters* **82**, pp. 1345–1349.

Brezger, B., Hackermüller, L., Uttenthaler, S., Petschinka, J., Arndt, M. and Zeilinger, A. (2002). *Physical Review Letters* **88**, 100404.

Busch, P. (2003). *Physical Review Letters* **91**, 120403.

Cassidy, D., Holton, G. and Rutherford, J. (2002). *Understanding Physics* (Springer), p. 632.

Caves, C., Fuchs, C.A., Manne, K.K. and Renes, J.M. (2004). *Foundations of Physics* **34**, pp. 193–209.

Cooke, R., Keane, M. and Moran, W. (1985). *Mathematical Proceedings of the Cambridge Philosophical Society* **98**, pp. 117–218.

Costella, J.P., McKellar, B.H.J. and Rawlinson, A.A. (1997). *American Journal of Physics* **65** (9), pp. 835–841.

Davies, P.C.W. (2007). *Cosmic Jackpot: Why Our Universe Is Just Right for Life* (Houghton Mifflin).

DeWitt, B.S. and Graham, R.N. (1971). *American Journal of Physics* **39**, pp. 724–738.

Dieks, D.G.B.J. (1996). *Communication & Cognition* **29** (2), pp. 153–168.

Dieks, D. (2001a). *International Studies in the Philosophy of Science* **15** (1), pp. 5–17.

Dieks, D. (2001b). *Studies in History and Philosophy of Modern Physics* **32** (2), pp. 217–241.

Duck, I. and Sudarshan, E.C.G. (1998). *American Journal of Physics* **66** (4), pp. 284–303.

Dutra, S.M. (2005). *Cavity Quantum Electrodynamics: The Strange Theory of Light in a Box* (Wiley Interscience).

Dyson, F.J. and Lenard, A. (1967/1968). *Journal of Mathematical Physics* **8**, pp. 423–434, and **9**, pp. 698–711.

Ehrenberg, W. and Siday, R.E. (1949). *Proceedings of the Physical Society B* **62**, pp. 8–21.

Einstein, A. (1948). *Dialectica* **2**, pp. 320–324.

Einstein, A. (1971). *The Born–Einstein Letters with Comments by Max Born* (Walker).

Einstein, A., Podolsky, B. and Rosen, N. (1935). *Physical Review* **47**, pp. 777–780; reprinted in Wheeler and Zurek (1983), pp. 138–141.

Elitzur, A.C. and Vaidman, L. (1993). *Foundations of Physics* **23**, pp. 987–997.

Englert, B.G., Scully, M.O. and Walther, H. (1994). *Scientific American* **271** (6), pp. 56–61.

Feynman, R.P. (1949). *Physical Review* **76**, pp. 749–759.

Feynman, R.P., Leighton, R.B. and Sands, M. (1963). *The Feynman Lectures in Physics*, Vol. 1 (Addison–Wesley), pp. 1–2.

Feynman, R.P., Leighton, R.B. and Sands, M. (1965). *The Feynman Lectures in Physics*, Vol. 3 (Addison–Wesley), Sec. 1–1.

Feynman, R.P. (1967). *The Character of Physical Law* (MIT Press), p. 129.

Feynman, R.P. (1985). *QED: The Strange Theory of Light and Matter* (Princeton University Press), p. 78.

Fodor, J. (1992). "The big idea: can there be a science of mind?" *Times Literary Supplement*, 3 July 1992.

Fuchs, C.A. (2001). In A. Gonis and P.E.A. Turchi (eds.), *Decoherence and its Implications in Quantum Computation and Information Transfer* (IOS Press), pp. 38–82.

Gleason, A.M. (1957). *Journal of Mathematics and Mechanics* **6**, pp. 885–894.

Goswami, A. (1995). *The Self–Aware Universe* (Tarcher).

Green, M.B., Schwarz, J.H. and Witten, E. (1988). *Superstring Theory* (Cambridge University Press).

Greenberger, D.M., Horne, M.A. and Zeilinger, A. (1989). In M. Kafatos (ed.), *Bell's Theorem, Quantum Theory, and Conceptions of the Universe* (Kluwer Academic), pp. 69–72.

Greenstein, G. and Zajonc, A. (1997). *The Quantum Challenge: Modern Research on the Foundations of Quantum Mechanics* (Jones and Bartlett).

Gribbin, J. and Rees, M. (1989). *Cosmic Coincidences* (Bantam Books).

Hanbury Brown, R. (1968). *Annual Reviews of Astronomy and Astrophysics 1968*, pp. 13–38.

Hanbury Brown, R., Davis, J. and Allen, L.R. (1967). *Monthly Notices of the Royal Astronomical Society* **137**, pp. 375–417.

Hanbury Brown, R. and Twiss, R.Q. (1956). *Nature* **178**, pp. 1046–1048.

Hardy, L. (1993). *Physical Review Letters* **71**, pp. 1665–1668.

Hawking, S. (1988). *A Brief History of Time* (Bantam), p. 125.

Hilgevoord, J. (1998). *American Journal of Physics* **66** (5), pp. 396–402.

Hoyle, F. (1959). *Religion and the Scientists* (SCM Press).

Jammer, M. (1974). *The Philosophy of Quantum Mechanics* (Wiley), pp. 68–69.

Jauch, J.M. (1968). *Foundations of Quantum Mechanics* (Addison–Wesley), pp. 92–94, 132.

Kaku, M. (1995). *Hyperspace* (Oxford University Press), p. 263.

Kochen, S. and Specker, E.P. (1967). *Journal of Mathematics and Mechanics* **17**, pp. 59–87.

Kolmogorov, A.N. (1950). *Foundations of the Theory of Probability* (Chelsea).

Laloë, F. (2001). *American Journal of Physics* **69** (6), pp. 655–701.

Landau, L.D. and Lifshitz, E.M. (1975). *A Course of Theoretical Physics*, Vol. 2: *The Classical Theory of Fields*, Fourth Revised English Edition (Elsevier Science), p. 266.

Lieb, E.H. (1976). *Reviews of Modern Physics* **48** (4), pp. 553–569.

Lieb, E.H. (2005). *The Stability of Matter: From Atoms to Stars—Selecta of Elliott H. Lieb* (Springer).

Lieb, E.H. and Seiringer, R. (2009). *The Stability of Matter in Quantum Mechanics* (Cambridge University Press).

Lieb, E.H. and Thirring, W.E. (1975). *Physical Review Letters* **35**, pp. 687–689, 1116.

Lockwood, M. (1989). *Mind, Brain and the Quantum* (Basil Blackwell).

London, F. and Bauer, E. (1939). Reprinted in Wheeler and Zurek (1983), pp. 217–259.

Marchildon, L. (2002). *Quantum Mechanics: From Basic Principles to Numerical Methods and Applications* (Springer).

du Marchie van Voorthuysen, E.H. (1996). *American Journal of Physics* **64** (12), pp. 1504–1507.

Marzke, R.F. and Wheeler, J.A. (1964). In H.-Y. Chiu and W.F. Hoffman (eds.) *Gravitation and Relativity* (W.A. Benjamin), p. 40.

Mattuck, R.D. (1976). *A Guide to Feynman Diagrams in the Many–Body Problem* (McGraw–Hill), p. 88.

McMahon, D. (2006). *Quantum Mechanics Demystified: A Self–Teaching Guide* (McGraw–Hill), pp. 186–188.

Mermin, N.D. (1985). Physics Today **38** (4), pp. 38–47.

Mermin, N.D. (1990). Physical Review Letters **65** (27), pp. 3373–3376.

Mermin, N.D. (1994). American Journal of Physics **62** (10), pp. 880–887.

Mermin, N.D. (2009). Physics Today **62** (5), pp. 8–9.

Misner, C.W., Thorne, K.S. and Wheeler, J.A. (1973). Gravitation (W.H. Freeman and Company), p. 1215.

Misra, B. and Sudarshan, E.C.G. (1977). Journal of Mathematical Physics **18**, pp. 756–763.

Nairz, O., Brezger, B., Arndt, M. and Zeilinger, A. (2001). Physical Review Letters **87**, 160401.

Newton, I. (1729). The Mathematical Principles of Natural Philosophy: Translated into English by Andrew Motte (Benjamin Motte).

Newton, T.D. and Wigner, E.P. (1949). Reviews of Modern Physics **21**, pp. 400–406.

Pauli, W. (1940). Physical Review **58**, pp. 716–722.

Pavičić, M., Merlet, J.-P., McKay, B. and McGill, N.D. (2005). Journal of Physics A **38**, pp. 1577–1592.

Penrose, R. (1986). In R. Penrose and C.J. Isham (eds.), Quantum Concepts in Space and Time (Clarendon Press), p. 139.

Penrose, R. (2005). The Road to Reality: A Complete Guide to the Laws of the Universe (Knopf).

Peres, A. (1980). American Journal of Physics **48**, pp. 931–932.

Peres, A. (1984). American Journal of Physics **52**, pp. 644–650.

Peres, A. (1995). Quantum Theory: Concepts and Methods (Kluwer).

Petersen, A. (1968). Quantum Physics and the Philosophical Tradition (MIT Press).

Phillips, S. (1995). Classical Indian Metaphysics (Open Court).

Pitowsky, I. (1998). Journal of Mathematical Physics **39**, pp. 218–228.

Primas, H. (2003). Mind and Matter **1** (1), pp. 81–119.

Radin, D. (2006). Entangled Minds: Extrasensory Experiences in a Quantum Reality (Paraview).

Redhead, M. (1987). Incompleteness, Nonlocality and Realism (Clarendon).

Rényi, A. (1955). Acta Mathematica Academia Scientiarum Hungaricae **6**, pp. 285–335.

Rényi, A. (1970). Foundations of Probability (Holden–Day).

Rohrlich, F. (1965). Classical Charged Particles (Addison–Wesley), pp. 65–66.

Rund, H. (1969). The Differential Geometry of Finsler Spaces (Springer).

Scully, M.O., Englert, B.G. and Walther, H. (1991). Nature **351** (6322), pp. 111–116.

Singh, I. and Whitaker, M.A.B. (1982). American Journal of Physics **50**, pp. 882–887.

Squires, E. (1990). Conscious Mind in the Physical World (Adam Hilger).

Sri Aurobindo (2001). Kena and Other Upanishads (Sri Aurobindo Ashram Publication Department).

Sri Aurobindo (2003). Isha Upanishad (Sri Aurobindo Ashram Publication Department).

Sri Aurobindo (2005). *The Life Divine* (Sri Aurobindo Ashram Publication Department).

Stapp, H.P. (2001). *Foundations of Physics* **31**, pp. 1465–1499.

Ulfbeck, O. and Bohr, A. (2001). *Foundations of Physics* **31** (5), pp. 757–774.

Vaidman, L. (1999). *Foundations of Physics* **29**, pp. 615–630.

von Neumann, J. (1955). *Mathematical Foundations of Quantum Mechanics* (Princeton University Press).

Weinberg, S. (1996). *The Quantum Theory of Fields*, Vol. 1 (Cambridge University Press), pp. 49–50.

Wheeler, J.A. (1983). In Wheeler and Zurek (1983), pp. 182–213.

Wheeler, J.A. and Zurek, W.H. (eds.) (1983). *Quantum Theory and Measurement* (Princeton University Press).

Whitehead, A.N. (1997/1925). *Science and the Modern World* (Free Press, Simon & Schuster), p. 51.

Wigner, E.P. (1961). In I.J. Good (ed.), *The Scientist Speculates* (Heinemann), pp. 284–302.

Wigner, E.P. (1959). *Group Theory* (Academic Press).

Wilczek, F. (1999). *Reviews of Modern Physics* **71** (2), S85–95.

Wilczek, F. (2001). *International Journal of Modern Physics A* **16**, pp. 1653–78.

Zee, A. (2003). *Quantum Field Theory in a Nutshell* (Princeton University Press).

Zurek, W.H. and Paz, J.P. (1997). In D.H. Feng and B.L. Hu (eds.), *Quantum–Classical Correspondence: Proceedings of the 4th Drexel Symposium on Quantum Nonintegrability* (International Press), pp. 367–380.

Index